Bioluminescence - Technology and Biology

Edited by Hirobumi Suzuki and Katsunori Ogoh

Published in London, United Kingdom

IntechOpen

Supporting open minds since 2005

Bioluminescence - Technology and Biology
http://dx.doi.org/10.5772/intechopen.87458
Edited by Hirobumi Suzuki and Katsunori Ogoh

Contributors
Ryo Nishihara, Kazuki Niwa, Ryoji Kurita, Tatsunosuke Tomita, Anielle C.A. Almeida Silva, Jerusa M. Maria de Oliveira, Kelen Da Silva, Francisco R.A. dos Santos, Rose Avelino, Marcelo Duzzioni, João Paulo Santos De Carvalho, Eurípedes A. da Silva Filho, Edigar Dias, Fábio De Oliveira, Malu Santos, Juliana E Silva, Marcos Silva, Carlo De Oliveira, Virmondes Rodrigues, Lucas Anhezini, Noelio Oliveira Dantas, kalpna Katiyar, Aditya Srivastava, Anchana Thancharoen, Shojiro A. Maki, Ryohei Saito-Moriya, Rika Obata, Hirobumi Suzuki, Katsunori Ogoh, Taro Hayashi, Yuichi Oba, Jeremy Mirza

First published in London, United Kingdom, 2021 by IntechOpen
IntechOpen is the global imprint of INTECHOPEN LIMITED, registered in England and Wales, registration number: 11086078, 5 Princes Gate Court, London, SW7 2QJ, United Kingdom
Printed in Croatia

British Library Cataloguing-in-Publication Data
A catalogue record for this book is available from the British Library

Additional hard and PDF copies can be obtained from orders@intechopen.com

Bioluminescence - Technology and Biology
Edited by Hirobumi Suzuki and Katsunori Ogoh
p. cm.
Print ISBN 978-1-83962-385-1
Online ISBN 978-1-83962-386-8
eBook (PDF) ISBN 978-1-83969-278-9

For EU product safety concerns:
IN TECH d.o.o., Prolaz Marije Krucifikse Kozulić 3, 51000 Rijeka, Croatia, info@intechopen.com or visit our website at intechopen.com.

Meet the editors

Dr. Hirobumi Suzuki received his Ph.D. in 1997 from Tokyo Metropolitan University, Japan, where he studied firefly phylogeny and the evolution of mating systems. He is especially interested in the genetic differentiation pattern and speciation process that correlate to the flashing pattern and mating behavior of some fireflies in Japan. He then worked for Olympus Corporation, a Japanese manufacturer of optics and imaging products, where he was involved in the development of luminescence technology and produced a bioluminescence microscope that is currently being used for gene expression analysis in chronobiology, neurobiology, and developmental biology. Dr. Suzuki currently serves as a visiting researcher at Kogakuin University, Japan, and also a vice president of the Japan Firefly Society.

Dr. Katsunori Ogoh received his Ph.D. from Shizuoka University, Japan, in 2005 for his study on sea-firefly. He is a member of the Research and Applications team of Olympus, specializing in biological techniques, and works in the sales and marketing department of microscopes, conducting research and development on the use of advanced analytical systems and applications, including cell culture, gene cloning, and microscopy applications. Dr. Ogoh has a powerful motivation for research and is working on bioluminescence technology development with his positive thinking and tireless efforts. He is inspired by diving in the sea. In his free time, he likes to go camping and make DIYs, and plays Splatoon video game with his family.

Contents

Preface

In recent years, luminescence technology, as well as fluorescence technology, has been applied in the medical and life science research fields for the measurement of biomolecules and for imaging *in vivo* and at the cellular level using luminescent markers as probes. The major difference between luminescence and fluorescence is the source of energy that excites the photon emitter, whether it is produced from a chemical reaction or from light. Therefore, luminescence measurement eliminates the problems of increased background due to autofluorescence and phototoxicity to living organisms in fluorescence measurement. However, fluorescence is superior in terms of emission intensity and image resolution. Now, with the development of luminescent and fluorescent materials and measurement techniques, the two measurement methods are developing as complementary technologies.

Luminescence is a phenomenon that can be interpreted in both physical and chemical terms. *Bioluminescence* in living organisms also falls into the category of chemical luminescence, since it is a phenomenon caused by chemical reactions occurring within the organism. The title of this book is *Bioluminescence* because it deals with luminescence of organisms mainly from the biological aspect, but of course it also includes the physical, chemical, and technical areas of luminescence. Following the previous series *Bioluminescence*: *Analytical Application and Basic Biology*, this book contains reviews and original articles in the fields of technology and biology.

Section 1 is focused on *Bioluminescence Techniques* contains four chapters. Chapter 1 contains a review article on fluorescent reporter proteins for monitoring the redox state of *Drosophila* cells and the application of fluorescent nanocrystals in cell lines. Chapter 2 contains a review article on imidazopyrazinone-type luciferin analogs that are specifically luminescent in the presence of human serum albumin rather than luciferase derived from luminescent organisms. The properties of the substrate and luminescence system for measuring low levels of human serum albumin as a biomarker for several diseases were presented. Chapter 3 contains a review article on firefly luciferin analogs that emit red-shifted near-infrared light. The new luciferin analog was combined with a mutant luciferase optimized for the analog and applied to *in vivo* imaging of mouse and marmoset tissues. The result was 10- to 1,400-fold increase in luminescence intensity, enabling detection of single-cell signals from mouse lungs and real-time imaging of marmoset brains at video rates. Chapter 4 is an original article in which the promoter activity of the adenylyl cyclase A gene in the fruiting body formation of the cellular slime mold *Dictyostelium* was analyzed using dual-color *bioluminescence* imaging techniques. By using two brightly modified firefly luciferases (green and red), the two promoter activities (proximal and distal regions) during the formation process from cell to fruiting body were imaged without the influence of autofluorescence.

Section 2 *Bioluminescent Organisms* contains three chapters. Chapter 5 is both a review and an original article. First, the concepts and practices of animal conservation and insect translocation are reviewed, and an original case study of a firefly translocation program in Thailand based on flashing patterns, mating behavior, and genetic population structure analyses is presented. Chapter 6 contains a review article on the origin of luciferin in marine luminescent organisms. Although the biosynthetic pathway of luciferin is not well understood, only a few luciferins of copepod and cyprid are known to be synthesized intrinsically from amino acids. On the other hand, most marine organisms acquire luciferin by preying on copepods and cyprid that make luciferin intrinsically. Chapter 7 is a review article on the ecology and behavior of bioluminescent organisms distributed from sea to land and taxonomically arranged from bacteria to vertebrates (fish). Almost all phyla are covered, and recent topics on new bioluminescent responses in fungi are also presented.

We would like to express our appreciation to the chapter authors for their outstanding work and also to Ms. Dolores Kuzelj, the Author Service Manager at IntechOpen, for effective communication and assistance during the preparation of this book.

Hirobumi Suzuki
Kogakuin University,
Tokyo, Japan

Katsunori Ogoh
Olympus Corporation,
Tokyo, Japan

Section 1

Bioluminescence Techniques

Chapter 1

Fluorescent Markers: Proteins and Nanocrystals

Anielle Christine Almeida Silva, Jerusa Maria de Oliveira, Kelen Talita Romão da Silva, Francisco Rubens Alves dos Santos, João Paulo Santos de Carvalho, Rose Kethelyn Souza Avelino, Eurípedes Alves da Silva Filho, Marcelo Duzzioni, Edigar Henrique Vaz Dias, Fábio de Oliveira, Juliana Rodrigues Machado, Malu Mateus Santos, Marcos Vinícius da Silva, Carlo José Freire de Oliveira, Virmondes Rodrigues Junior, Lucas Anhezini and Noelio Oliveira Dantas

Abstract

This book chapter will comment on fluorescent reporter proteins and nanocrystals' applicability as fluorescent markers. Fluorescent reporter proteins in the Drosophila model system offer a degree of specificity that allows monitoring cellular and biochemical phenomena *in vivo*, such as autophagy, mitophagy, and changes in the redox state of cells. Titanium dioxide (TiO_2) nanocrystals (NCs) have several biological applications and emit in the ultraviolet, with doping of europium ions can be visualized in the red luminescence. Therefore, it is possible to monitor nanocrystals in biological systems using different emission channels. CdSe/CdS magic-sized quantum dots (MSQDs) show high luminescence stability in biological systems and can be bioconjugated with biological molecules. Therefore, this chapter will show exciting results of the group using fluorescent proteins and nanocrystals in biological systems.

Keywords: nanocrystals, fluorescent proteins, fluorescent markers, magic-sized quantum dots, titanium dioxide

1. Introduction

Several types of tools have been developed in order to monitor biological processes through fluorescence images. Some of these tools are the use of fluorescent proteins and nanomaterials. This book chapter will comment in particular on green fluorescent protein and luminescent nanocrystals.

The green fluorescent protein (GFP) of Jellyfish *Aequoria victoria* and fluorescent homologous proteins of different colors isolated from other sea creatures have led to the development of fluorophores that have been widely used in recent decades. In the biological area, fluorescent antibodies are a powerful tool for analyzing the subcellular location of proteins of interest. In addition, a gene encoding a fluorescent protein can be introduced into a model organism, resulting in the expression of functional fluorescent proteins, which can then be detected by fluorescence microscopy, flow cytometry, and other fluorescence-based methods. Fluorescent proteins have revolutionized biological and biomedical research, for example, it is possible to monitor the activity of a gene promoter by placing a fluorescent protein under its control. By using this tool, the spatial and temporal patterns of gene activation were revealed, as well as it was possible to trace the specific fates of cell populations or even to visualize the different shapes that cells can assume during development. Perhaps the most sophisticated area is the construction of genetically modified fluorescent sensors for the detection of ions, small molecules, and various types of enzyme activity [1–3].

One of the best and most used *in vivo* models to investigate biological phenomena is the fruit fly. The fruit fly *Drosophila melanogaster* is a well-established model organism in nanotoxicology studies [4]. *Drosophila* has a short life cycle, low maintenance cost, and a considerable amount of conserved genes and physiological mechanisms with humans [5, 6]. The complete sequencing of the *Drosophila* genome combined with genetic editing techniques allows the construction of reporter lines (for example, GFP) fused to specific genes [6]. One of the main tools that make the fruit fly an excellent model organism is the possibility of expressing genes of interest in specific tissues through the UAS-GAL4 binary expression system. This system consists of two factors: the GAL4 transcription factor fused to the promoter region of a gene of interest and the upstream activation sequence (UAS). GAL4 is able to bind the upstream activation sequence (UAS), activating the transcription of a target gene linked to UAS, allowing ectopic gene expression [6]. Thus, the use of fluorescent reporter genes under the control of the UAS-GAL4 system allows a degree of specificity necessary to monitor cellular and biochemical phenomena *in vivo* in different tissues, such as autophagy, [7] mitophagy [8], and changes in the redox state of cells [1]. Taken together, these features are essential for studies that evaluate the effects of xenobiotics on development, however, it can still be improved in the nanotoxicology area, mainly in the use of reporter lines for the elucidation of cellular mechanisms responsible for toxicity and the subcellular localization of nanocrystals.

The development of different nanoscale materials has increased for different applications. Titanium dioxide (TiO_2) nanocrystals (NCs) have been used in several types of cosmetics, food, and the textile industry [9, 10]. This is because this NC has a wide variety of properties that improve materials, such as its bioluminescence and chemical stability [11]. Bioluminescent techniques are widely used in biomedicine for studies of drug screening, molecular markers, and monitoring of molecular reactions, among other applications [12]. Bioluminescent NCs, such as TiO2, present an excellent opportunity to obtain ultra-sensitive and enhanced analyzes and images, in addition to allowing the study of bioluminescence [13–15]. The use of bioluminescent imaging in vivo allows the visualization of biological processes in intact living organisms, providing abundant quantitative space–time information beyond the reach of conventional in vitro tests and fixed material [15].

Doping is a technique that allows the incorporation of substitutional ions into the crystalline structure of materials, generating exciting properties [16]. TiO_2 nanocrystals (NCs) with europium ions incorporated in their structure can be visualized in red fluorescence [17]. This acquired property makes it possible to track luminescence, thus being able to be coupled to biomolecules and drugs for studies of effects and tracking them, for example, which can assist in the studies of quantitative monitoring of molecular reactions and cellular behaviors, allowing a better understanding of the functions dynamic and complicated biological phenomena [18, 19].

Quantum dots (QDs) of cadmium chalcogenides (CdSe, CdS, and CdTe) absorb and emit in the visible electromagnetic spectrum, and for this reason, they are used in several applications of biological and biomedical marking, such as fluorescent probes, biosensors, and others. In the area of biological labeling, the great applicability of QDs occurs because they present several advantages over traditional organic fluorophores, such as a long fluorescence life span, ~ 100 times greater, which allows to distinguish it from the background b signal, seen that autofluorescence has a much shorter fluorescence life; absorption and emission spectra tunable; high photo resistance and chemo-degradation; and high fluorescence intensity [20–22]. However, this comparison of the fluorescence intensity of the QDs was performed in non-aqueous solvents, with unconjugated QDs, and in non-biological media, since the fluorescence intensity may be lower when the QDs are conjugated and used in biological labeling experiments [23].

Ultra-small PQs (USPQs) are nanocrystals with extremely small sizes, presenting strong quantum confinement effects, in which most of their atoms are located on the surface [24]. A large number of atoms on the surface and the presence of several pendant bonds lead to changes in the properties of nanocrystals, which can be observed in the fluorescence spectra [25].

The quantum dots of magic-sized (MSQDs) are nanocrystals with extremely small sizes (<2 nm) and that present physical property utterly different from traditional QDs [26]. Although MSQDs have similar properties to USQDs, including composition and size, some fundamental properties place these QDs in different classes. The characteristic properties of MSQDs are thermodynamically stable structures, wide luminescence range, high size stability over time, relatively narrow absorption spectra and/or heterogeneous (discontinuous) growth [27–31]. The structures are thermodynamically stable; they are formed from the arrangement of a certain number of atoms, which gives it high stability. Nguyen et al. made theoretical predictions of different types of CdSe MSQDs structures aligned with the literature's experimental results [32]. The term magic size is related to a (magic) number of atoms in the structure that makes QDs extremely stable [32]. The broad luminescence spectrum occurs due to MSQDs having internal atomic defects (absence or extra presence of atoms) [27, 29, 32].

The development of new alternatives for the study of biomolecules in organic systems has grown considerably. The high specificity and sensitivity of scientific methodologies based on fluorescence clarify biological events [33]. Fluorescent probes based on organic dyes have been shown to identify biomolecules [34, 35]. Silva et al. demonstrated that the biocompatibility of CdSe/CdS MSQRd could be tuned in the synthesis, [36] present high luminescence stability in biological systems [37], can be bioconjugated with several biomolecules aiming at the most diverse luminescent probes [38–42] and in biosensors [43, 44].

This chapter shows recent results that the group has been working with fluorescent reporter proteins and the applicability of nanocrystals as fluorescent markers. Nanocrystals of pure and europium doped TiO_2 and CdSe/CdS (MSQDs) will be some of the exciting tools for marking in biological systems.

2. Fluorescent proteins and nanocrystals

This section will show the group results using GFP tagged proteins and nanocrystals' applicability as fluorescent markers.

2.1 Drosophila lines expressing fluorescent proteins

In 2011, Albrecht et al. established a monitoring system that allows assessing the status of chemically defined redox species (the redox pair GSH/GSSH and H_2O_2) in subcellular compartments cytosol and mitochondria *in vivo*. They have fused a probe sensitive to redox changes (ro-GFP2) [45–47] to the microbial H_2O_2 sensor oxidant receptor peroxidase 1 (Orp1) [46]. In a reduced state, this probe exhibits excitation around 488 nm, while upon oxidation, roGFP2 gains excitability at 405 nm and loses excitability at 488 nm. In the present work, we used one of the transgenic *Drosophila* lines described by Albrecht and collaborators, called mito-roGFP2-Orp1 [1] to exemplify how *in vivo* sensors can be valuable for analyzing the redox state and to propose its use for the analysis of nanomaterials biocompatibility *in vivo*. In **Figure 1** we show different dissected larval tissues of *Drosophila*

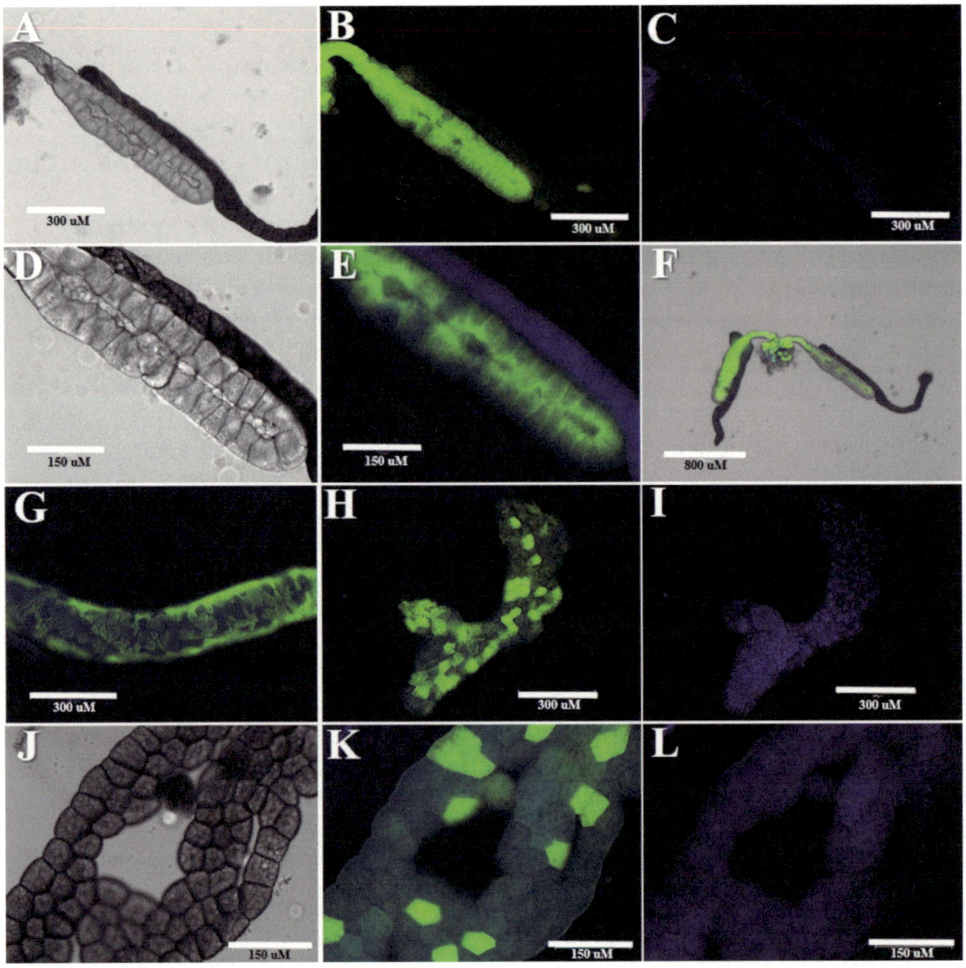

Figure 1.
Drosophila *lines expressing fluorescent proteins can be used as* in vivo *sensors of redox status. Different GAL4 drivers were used to express Mito-orp1-GFP in different* Drosophila *larval tissues. (A-F) The* Drosophila *GAL4 driver sgs3-GAL4 was used to express Mito-orp1-GFP2 in larval salivary glands. A- bright field image of a* Drosophila *larval salivary gland. In (B) the Mito-orp1-GFP can be visualized in its reduced state, while in (C) a weak signal is seen under 405 nm light. A greater magnification of the salivary gland shown in a can be seen in (D) while in (E) the overlap of the Mito-orp1-GFP2 in its reduced (488 nm) and oxidized (405 nm) state is shown. (G) Larval midgut showing the overlap of the Mito-orp1-GFP2 in its reduced (488 nm) and oxidized (405 nm) state. The Mito-orp1-GFP2 in its reduced state is shown in (H) and (K) while the sensor oxidized fluorescence is seen in (I) and (L). All images were acquired using ThermoFisher Scientific EVOS M7000 Imaging System.*

expressing the redox sensor mito-roGFP2-Orp1. **Figure 1A** shows the bright field image of a dissected larval salivary gland, while in 3B we can see that there is a high concentration of mito-roGFP2-Orp1 in its reduced state, evidenced by green fluorescence, and a low concentration of mito-roGFP2-Orp1 in its oxidized state. In 1D, a higher magnification image of the salivary gland in 1A is shown, while in 1E the merged image of 488 nm and 405 nm channels shows the balance between reduced and oxidized mito-roGFP2-Orp1. In 1G, a portion of the midgut also shows a higher concentration of reduced mito-roGFP2-Orp1. **Figure 1H** shows reduced mito-roGFP2-Orp1 distribution throughout the larval fat body. It is interesting to notice that there is a clear difference in the concentration of reduced mito-roGFP2-Orp1 in cells within the same tissue, which is even more evident in the image in greater magnification shown in 1 K. The samples in 1I and 1 L show the oxidized mito-roGFP2-Orp1 in the same larval fat body. As expected, our analysis of control samples showed that most mito-roGFP2-Orp1 proteins are in its reduced state, exhibiting excitation around 488 nm. We are currently using this valuable tool to analyze the effect of different nanocrystals on the redox balance in *Drosophila* as an additional approach for the determination of biocompatibility *in vivo*.

Figure 2 shows three different transgenic lines of *Drosophila* that can be used to assist in the subcellular localization of fluorescent nanoparticles. **Figure 2A-C** shows dissected tissues of the D1-GFP transgenic line (BL.66454). The D1-GFP protein binds to chromosomes allowing the nuclei visualization. In **Figure 2A** we can see a pair of larval salivary glands while **Figure 2B-C** shows different portions of the larval gut. The progeny of the cross between the lineage mef2-Gal4 and UAS-mito-GFP (BL. 8443) was used to visualize the larval muscles (**Figure 2D-F**). This is because mef2-Gal4 drives Gal4 expression in muscles where it binds to the regulatory sequence UAS-mito-GFP, which in turn regulates the expression of a

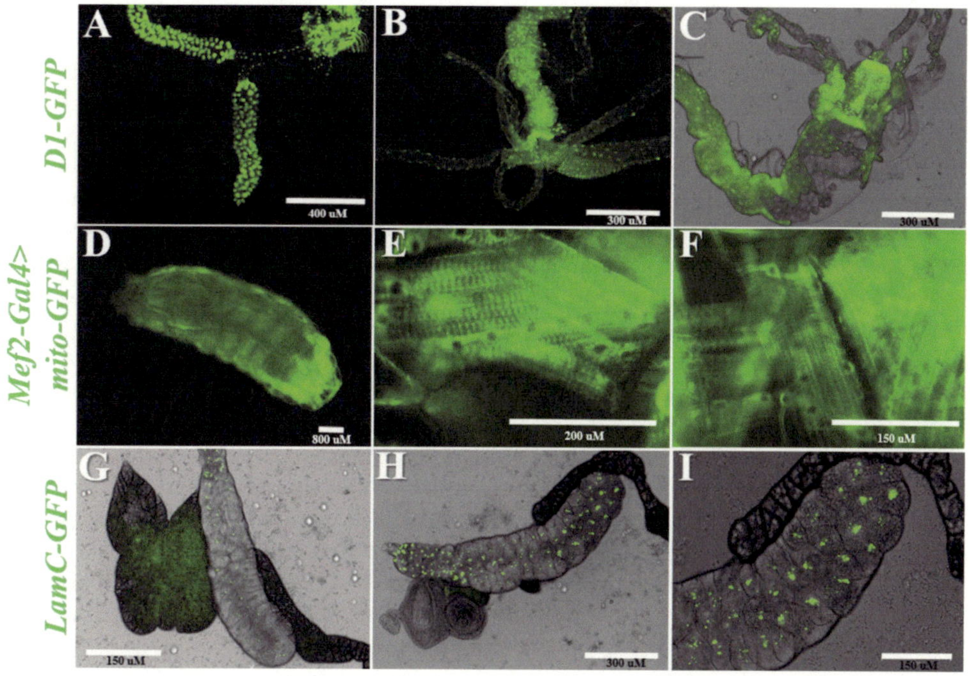

Figure 2.
The expression of fluorescent proteins in Drosophila *as a tool to visualize cellular subcompartments. (A-C) D1-GFP expression in* Drosophila *larval tissues. A pair of larval salivary glands is shown in a while (B) and (C) shows different portions of the larval gut. (D-F) L3 larvae expressing Mito-GFP in muscles. (G-I) shows salivary glands expressing LamC-GFP localization at the nuclear envelope of cells.*

mitochondrial import sequence fused to the fluorescent protein GFP, allowing the visualization of mitochondria in the muscle tissues. **Figure 2G-I** shows salivary glands of the LamC-GFP transgenic lineage (BL. 6837) which allows for the visualization of the nuclear envelope of cells. In these examples, we showed 3 different *Drosophila* transgenic lines in the green band (GFP) that can be used to assist in the subcellular localization of fluorescent nanoparticles, however, it is essential to notice that hundreds of lines are available. Data on the genome and the wide range of reporter lines of *Drosophila* can be found on Flybase (https://flybase.org/). Reporter lines of *Drosophila* can be purchased in collections such as Bloomington *Drosophila* Stock Center (BDSC) at Indiana University. These different tools allow the investigators to choose from a great variety of cell types, subcellular compartments as well as the fluorescence band that best adapts to the nanoparticles of interest.

2.2 Nanocrystals as luminescent markers (nanomarkers)

Figure 3 shows exciting results on pure and europium (Eu) doped TiO_2 NCs. TiO_2 NCs absorb and emit in the ultraviolet, but when incorporating the europium ions in its crystalline structure, by replacing some titanium ions, it shows luminescence in red. The colors emitted by the pure and Eu doped TiO_2 NCs (**Figure 3a**), and the crystalline structure in the anatase phase (**Figure 3b**) are illustrated. Also, in **Figure 3c**, the emission spectra of these nanocrystals are observed.

In order to investigate whether TiO_2 and TiO_2:Eu nanocrystals could be tracked on adult *Drosophila* after exposure during development TiO_2 and TiO_2:Eu nanocrystals were mixed in standard *Drosophila* culture medium at the final concentration of 100 mM.. The larvae were carefully staged and transferred as L1 (first instar larvae) to medium containing TiO_2 and TiO_2:Eu. The control contained only a standard *Drosophila* culture medium. The animals developed through all larval stages during the following 3 days. At this stage, the larvae actively feed until they become pupae. After pupal metamorphosis, the animals emerged as adults were dissected and its abdominal fat body was analyzed through fluorescence microscopy under UV light to analyze the TiO_2 bioaccumulation and under red light to detect TiO_2:Eu. All samples images were acquired using the same light intensity and exposure time. **Figure 4** shows the tracking data of TiO_2 and TiO_2:Eu in the fat body of adult animals after exposure during the larval stage. It is possible to observe that the fat body spheres of the control animals (**Figure 4A** and **C**) show intrinsic fluorescence when excited with ultraviolet light, however when the animals were exposed to TiO_2 the intensity of fluorescence was significantly higher (**Figure 4B** and **D**).

In order to distinguish between intrinsic fluorescence from fat body and TiO_2 fluorescence, the pixel intensity was measured and compared among all fat body spheres of control images and TiO_2 treated samples. As we can observe in the graphic in **Figure 5a** there was a drastic increase in fluorescence due to the presence of TiO_2. The fat body spheres of the control animals (**Figure 4E** and **G**) also showed intrinsic fluorescence when excited with red light; however, when the animals have exposed to TiO_2:Eu the intensity of fluorescence was higher (**Figure 4F** and **H**). The pixel intensity analysis showed that the presence of TiO_2:Eu caused a significant increase in fluorescence (**Figure 5b**). The observation that the NCs of TiO_2 and TiO_2:Eu could be detected in the fat body of newly emerged adult animals indicates that the bioaccumulation of nanocrystals during larval development persisted until the beginning of the adult stage. Surprisingly, we observed that animals dissected on the second day of its emergence no longer had fat bodies fluorescent spheres containing nanocrystals. This may indicate that one day following the emergence, the animals were able to excrete the NC. The disappearance of nanocrystals may

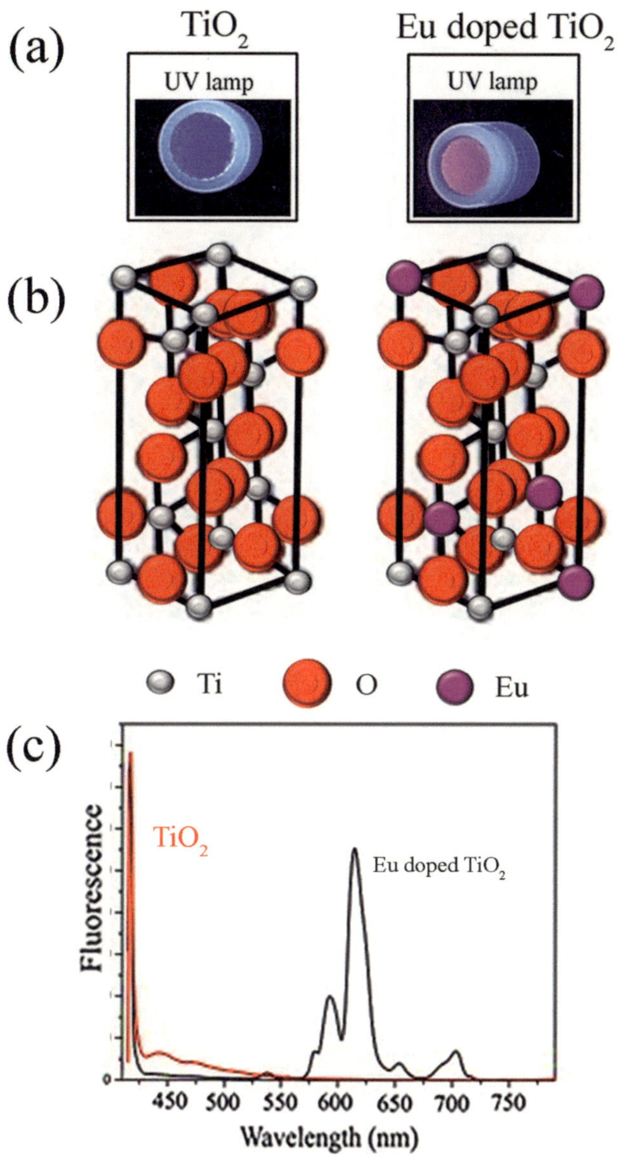

Figure 3.
(a) Photographic image of nanopowders, (b) anatase crystalline structure, (c) luminescence spectra of pure and Eu doped TiO_2 NCs.

also be related to the rapid absorption of the fat body during the first days of life. Similar results were described by Jovanovic et al. 2016, which observed that animals that received TiO_2 during the larval stage did not have TiO_2 as adults [9].

The optical properties and illustration of CdSe/CdS are shown in **Figure 6**. The aqueous solution and the illustration of the core/shell structure of CdSe/CdS MSQDs with a surfactant are exemplified to facilitate understanding (**Figure 6a, b**). The optical absorption and broad luminescence spectra are characteristics of magic-sized quantum dots of CdSe/CdS (**Figure 6c**). In addition, one of the essential properties of the CdSe/CdS MSQDs that allows its application in biological systems is entering and staying inside cells. To test this capacity, we incubated a classical macrophage cell line (RAW 264.7) with CdSe/CdS MSQDs of (200 ng/μL) and evaluated their internalization by Flow Cytometry in different time points (1 to 60 minutes). Flow

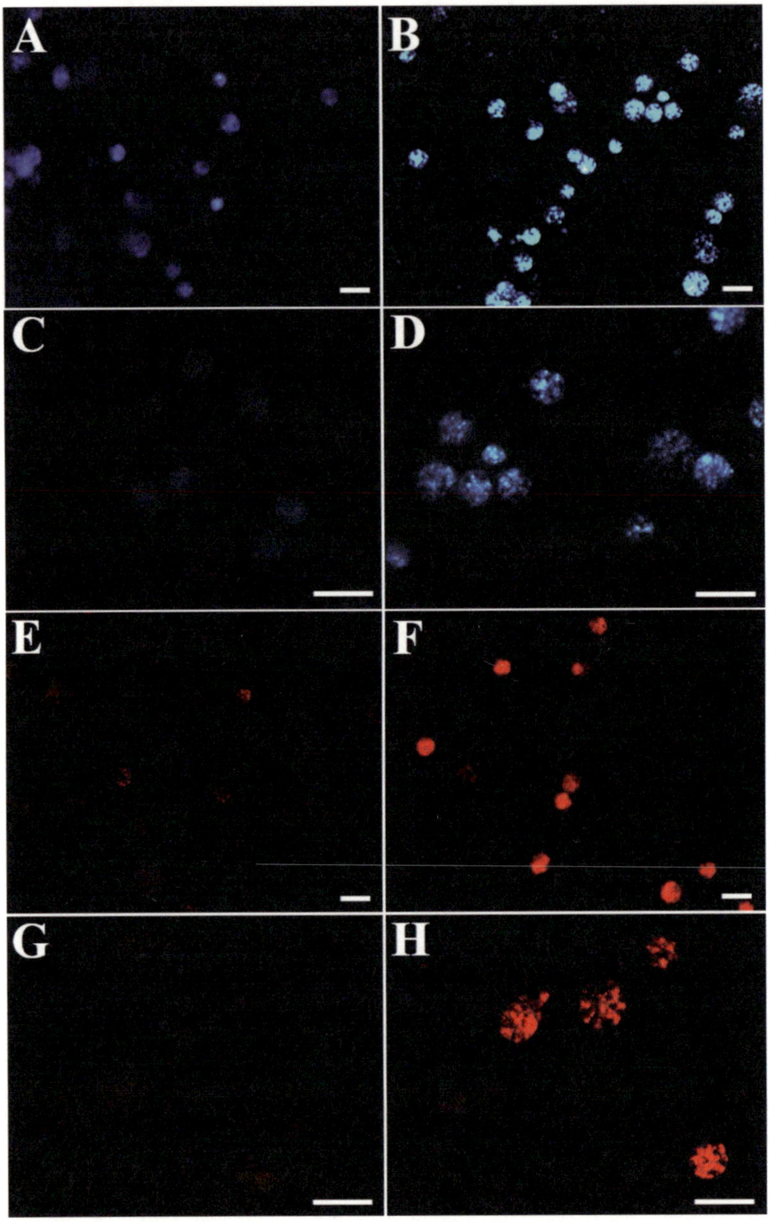

Figure 4.
In vivo *fluorescence of* TiO_2 *and* TiO_2*:Eu in adult* Drosophila *fat body.* Drosophila *tissues as the fat body shows a well-known intrinsic fluorescence as observed in representative images of control animals (A) and (C) at 405 nm E and G at 546 nm), however in the* TiO_2 *(B and D) and* TiO_2*:Eu (F and H) treated animals it is possible to observe a drastic increase in luminescence when compared to control. Scale bar represents 50uM.*

Cytometry is a unique methodological approach to determine cell staining as it evaluates a considerable number of cells per second, one by one, and reports if cells are fluorescent. Just after 1 minute, the MSQDs nearly 50% of cells were fluorescent, and this percentual was growing to >97% after 60 minutes (**Figure 6c, d**).

Bioimaging assays are biological applications QDs since they can be bioconjugated with proteins, antibodies, and DNA [39, 48, 49]. In general, these tests depend on the biocompatibility of QDs, which is obtained by functionalizing the surface of these nanoparticles [39, 50–52]. The bioconjugation allows the study and tracking of biomolecules in biological systems such as cell cultures

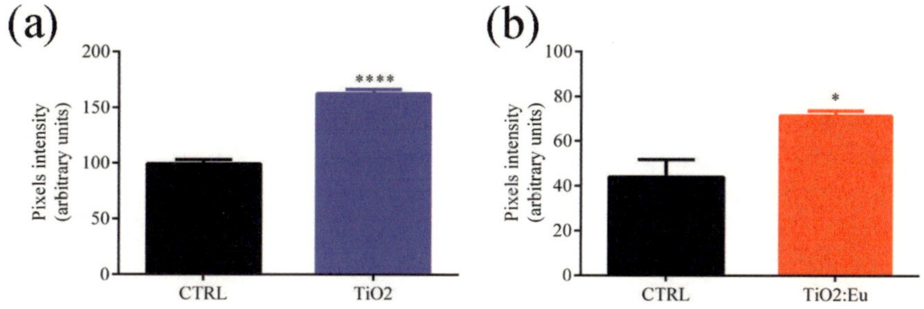

Figure 5.
Indirect quantification of TiO_2 and TiO_2:Eu fluorescence. (a) Pixel intensity analysis of fat body spheres of TiO_2 treated Drosophila *to control fat body spheres. (b) Pixel intensity analysis of fat body spheres of TiO_2:Eu treated animals compared to control fat body spheres.*

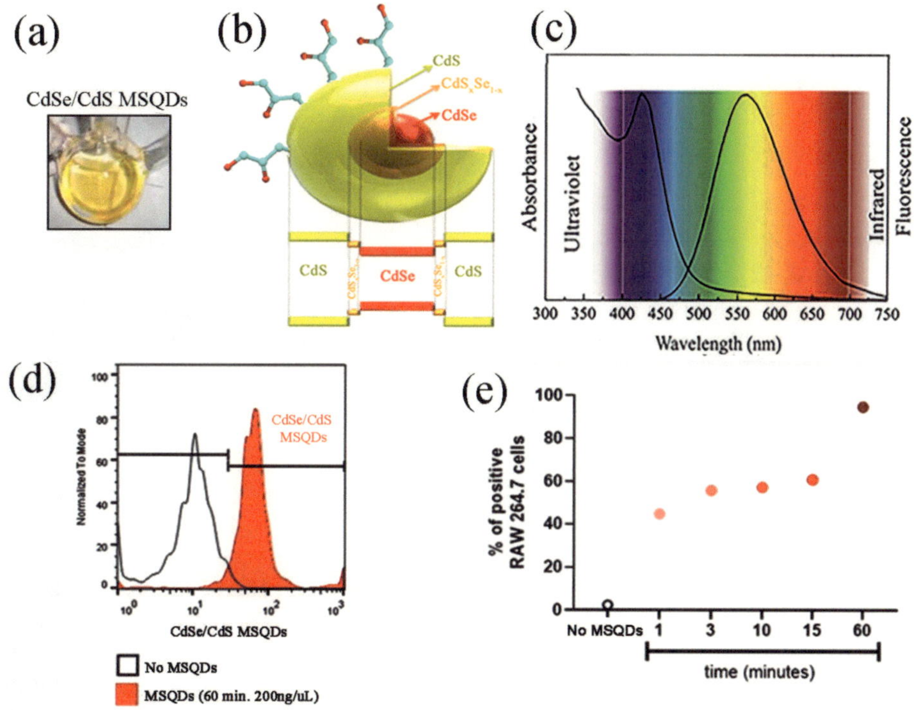

Figure 6.
(a) Photographic image of solution, (b) illustration of CdSe/CdS MSQDs, (c) optical absorption/luminescence spectra of CdSe/CdS MSQDs (d, e) incorporation of MSQDs by RAW 264.7. The intracellular location was determined by flow cytometry after incubation of CdSe/CdS MSQDs (200 ng/uL) with RAW 264.7 cell line (1x10^4/mL) at different time points. Cells were washed in saline solution before acquisition to exclude extracellular MSQDs. At least 5000 events were acquired in a FacsCalibur flow cytometer.

and laboratory animals [53, 54]. The versatility of QDs associated with maltose-binding protein for intracellular delivery of the drug beta-cyclodextrin [55]. Other studies have used the quantum dots for *in vivo* multiphoton biologic imaging. Kwon et al. conjugated iron selenide QDs with monoclonal human epidermal growth factor receptor 2 antibodies to study xenograft breast tumor model in mice [56].

The tracking and study of biomolecules labeled with QDs in vitro and in vivo is a reality in several areas, allowing us to analyze the location and distribution of bioconjugate in biological systems. Silva et al. demonstrated that the CdSe/CdS MSQDs could be bioconjugated with several biomolecules aiming at the most

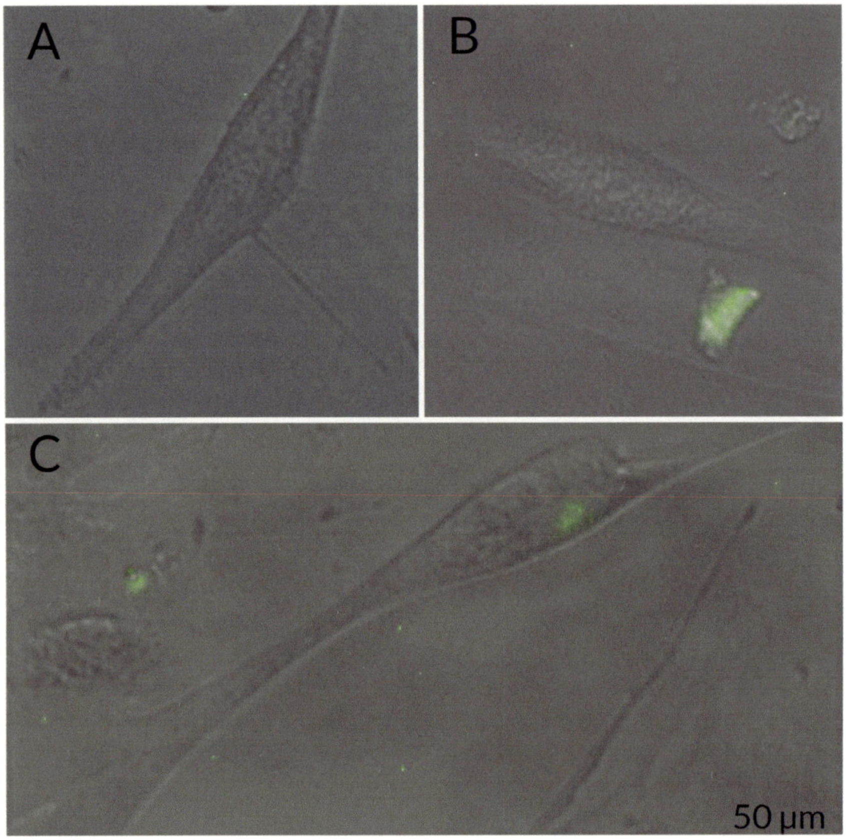

Figure 7.
Fluorescence microscopy images showing the tracking of BaltPLA$_2$ in myoblast culture. (A) Cell control (myoblasts only); (B) myoblasts treated with MSQDs (200 ng/µL) for 18 h; (C) myoblasts treated with MSQDs (200 ng/µL)-BaltPLA$_2$ (100 ng/µL) for 18 h. scale 50 µm.

diverse luminescent probes [38–42] in biosensors [43, 44]. Dias et al. labeled a phospholipase A2 isolated from *Bothrops alternatus* snake venom with CdSe/CdS MSQDs to track it in myoblast culture, making it possible to identify the bioconjugate on the surface of the plasma membrane and in the nuclear region [39]. **Figure 7** corroborates these data since it is possible to observe fluorescence markings only in myoblasts treated with the bioconjugate MSQDs-BaltPla2.

3. Conclusion

In this chapter, we have shown that fluorescent reporter proteins in the *Drosophila* model system are excellent tools to monitor cellular and biochemical phenomena *in vivo*, such as changes in the redox state of cells, as well as are a valuable tools to assist in the subcellular localization of fluorescent nanoparticles. We also showed that TiO_2 and Eu doped TiO_2 NCs fluorescence could be detected in adult animals following exposure during development. Intracellular location of CdSe/CdS MSQDs in RAW 264.7 cell line and tracking of BaltPLA$_2$ bioconjugated in myoblast culture. Therefore, the use of fluorescent proteins and nanocrystals *in vivo* are exciting tools as they provide abundant qualitative and quantitative data and allow the visualization of biological processes in intact cells and living organisms.

Acknowledgements

This work was supported by grants of CNPq, CAPES, FAPEAL, and FAPEMIG.

Conflict of interest

The authors declare no conflict of interest.

Author details

Anielle Christine Almeida Silva[1,2*], Jerusa Maria de Oliveira[3], Kelen Talita Romão da Silva[1,3], Francisco Rubens Alves dos Santos[1,3], João Paulo Santos de Carvalho[1], Rose Kethelyn Souza Avelino[3], Eurípedes Alves da Silva Filho[4], Marcelo Duzzioni[5], Edigar Henrique Vaz Dias[6], Fábio de Oliveira[7], Juliana Rodrigues Machado[8], Malu Mateus Santos[9], Marcos Vinícius da Silva[9], Carlo José Freire de Oliveira[9], Virmondes Rodrigues Junior[9], Lucas Anhezini[3] and Noelio Oliveira Dantas[1*]

1 Laboratory of New Nanostructured and Functional Materials, Physics Institute, Federal University of Alagoas, Maceió, Alagoas Brazil

2 Programa de Pós-Graduação da Rede Nordeste de Biotecnologia (RENORBIO), Federal University of Alagoas, Maceió, Alagoas Brazil

3 Laboratory for in vivo Toxicity, Institute of Biological Sciences and Health, Federal University of Alagoas, Maceió, Alagoas, Brazil

4 Laboratory of Genetic and Applied Microbiology, Institute of Biological Sciences and Health, Federal University of Alagoas, Maceió, Alagoas, Brazil

5 Laboratory of Pharmacological Innovation, Institute of Biological Sciences and Health, Federal University of Alagoas, Maceió, Alagoas, Brazil

6 State University of Minas Gerais, Ituiutaba, MG, Brazil

7 Institute of Biomedical Sciences, Federal University of Uberlândia, Uberlândia, MG, Brazil

8 Department of Pathology, Genetics and Evolution, Federal University of Triângulo Mineiro, Uberaba, Minas Gerais, Brazil

9 Department of Microbiology, Immunology, and Parasitology, Institute of Biological and Natural Sciences, Federal University of Triângulo Mineiro, Uberaba, MG, Brazil

*Address all correspondence to: aniellechristineas@gmail.com and noelio@fisufal.br

References

[1] Albrecht, S.C.; Barata, A.G.; Großhans, J.; Teleman, A.A.; Dick, T.P. In vivo mapping of hydrogen peroxide and oxidized glutathione reveals chemical and regional specificity of redox homeostasis. *Cell Metab.* 2011, *14*, 819-829, doi:10.1016/j.cmet.2011.10.010.

[2] Chudakov, D.M.; Matz, M. V.; Lukyanov, S.; Lukyanov, K.A. Fluorescent proteins and their applications in imaging living cells and tissues. *Physiol. Rev.* 2010, *90*, 1103-1163, doi:10.1152/physrev.00038.2009.

[3] VanEngelenburg, S.B.; Palmer, A.E. Fluorescent biosensors of protein function. *Curr. Opin. Chem. Biol.* 2008, *12*, 60-65, doi:10.1016/j.cbpa.2008.01.020.

[4] Ong, C.; Yung, L.Y.L.; Cai, Y.; Bay, B.H.; Baeg, G.H. Drosophila melanogaster as a model organism to study nanotoxicity. *Nanotoxicology* 2015, *9*, 396-403, doi:10.3109/17435390.2014.940405.

[5] Pandey, U.B.; Nichols, C.D. Human disease models in *Drosophila melanogaster* and the role of the fly in therapeutic drug discovery. *Pharmacol. Rev.* 2011, *63*, 411-436, doi:10.1124/pr.110.003293.

[6] Hales, K.G.; Korey, C.A.; Larracuente, A.M.; Roberts, D.M. Genetics on the fly: A primer on the drosophila model system. *Genetics* 2015, *201*, 815-842, doi:10.1534/genetics.115.183392.

[7] Devorkin, L.; Gorski, S.M. Monitoring autophagy in drosophila using fluorescent reporters in the UAS-GAL4 system. *Cold Spring Harb. Protoc.* 2014, *2014*, 967-972, doi:10.1101/pdb.prot080341.

[8] Kim, Y.Y.; Um, J.H.; Yoon, J.H.; Kim, H.; Lee, D.Y.; Lee, Y.J.; Jee, H.J.; Kim, Y.M.; Jang, J.S.; Jang, Y.G.; et al. Assessment of mitophagy in mt-Keima Drosophila revealed an essential role of the PINK1-Parkin pathway in mitophagy induction in vivo. *FASEB J.* 2019, *33*, 9742-9751, doi:10.1096/fj.201900073R.

[9] Jovanović, B.; Jovanović, N.; Cvetković, V.J.; Matić, S.; Stanić, S.; Whitley, E.M.; Mitrović, T.L. The effects of a human food additive, titanium dioxide nanoparticles E171, on *Drosophila melanogaster* - a 20 generation dietary exposure experiment. *Sci. Rep.* 2018, *8*, 1-12, doi:10.1038/s41598-018-36174-w.

[10] Chen, X.; Mao, S.S. Titanium dioxide nanomaterials: Synthesis, properties, modifications and applications. *Chem. Rev.* 2007, *107*, 2891-2959.

[11] Kamat, P. V. *Journal of Physical Chemistry C*. 2012, pp. 11849-11851.

[12] Fleiss, A.; Sarkisyan, K.S. A brief review of bioluminescent systems (2019). *Curr. Genet.* 2019, *65*, 877-882, doi:10.1007/s00294-019-00951-5.

[13] Duarte, C.A.; Goulart, L.R.; Filice, letícia de S.C.; De Lima, I.L.; Campos-Fernándes, E.; Dantas, N.O.; Silva, A.C.A.; Soares, M.B.P.; Santos, R.R.; CArdoso, C.M.A.; et al. Characterization of Crystalline Phase of TiO 2. *Materials (Basel).* 2020, *13*, 4071.

[14] Barkalina, N.; Charalambous, C.; Jones, C.; Coward, K. Nanotechnology in reproductive medicine: Emerging applications of nanomaterials. *Nanomedicine Nanotechnology, Biol. Med.* 2014, *10*, e921–e938, doi:10.1016/j.nano.2014.01.001.

[15] Padmanabhan, P.; Kumar, A.; Kumar, S.; Chaudhary, R.K.; Gulyás, B. Nanoparticles in practice for molecular-imaging applications: An overview. *Acta*

Biomater. 2016, *41*, 1-16, doi:10.1016/j.actbio.2016.06.003.

[16] Bharat, T.C.; Shubham; Mondal, S.; Gupta, H.S.; Singh, P.K.; Das, A.K. Synthesis of doped zinc oxide nanoparticles: A review. In Proceedings of the Materials Today: Proceedings; Elsevier Ltd, 2019; Vol. 11, pp. 767-775.

[17] Sandoval, S.; Yang, J.; Alfaro, J.G.; Liberman, A.; Makale, M.; Chiang, C.E.; Schuller, I.K.; Kummel, A.C.; Trogler, W.C. Europium-doped TiO 2 hollow nanoshells: Two-photon imaging of cell binding. *Chem. Mater.* 2012, *24*, 4222-4230, doi:10.1021/cm302642g.

[18] Ghaderi, S.; Ramesh, B.; Seifalian, A.M. Fluorescence nanoparticles "quantum dots" as drug delivery system and their toxicity: A review. *J. Drug Target.* 2011, *19*, 475-486, doi:10.3109/1061186X.2010.526227.

[19] Vollrath, A.; Schubert, S.; Schubert, U.S. Fluorescence imaging of cancer tissue based on metal-free polymeric nanoparticles-a review. *J. Mater. Chem. B* 2013, *1*, 1994-2007, doi:10.1039/c3tb20089b.

[20] Alivisatos, P. The use of nanocrystals in biological detection. *Nat. Biotechnol.* 2004, *22*, 47-52, doi:10.1038/nbt927.

[21] Resch-Genger, U.; Grabolle, M.; Cavaliere-Jaricot, S.; Nitschke, R.; Nann, T. Quantum dots versus organic dyes as fluorescent labels. *Nat Methods.* 2008, *5*, 763-75.

[22] Deerinck, T.J. The application of fluorescent quantum dots to confocal, multiphoton, and electron microscopic imaging. *Toxicol. Pathol.* 2008, *36*, 112-6, doi:10.1177/0192623307310950.

[23] Wu, X.; Liu, H.; Liu, J.; Haley, K.N.; Treadway, J.A.; Larson, J.P.; Ge, N.; Peale, F.; Bruchez, M.P. Immunofluorescent labeling of cancer marker Her2 and other cellular targets with semiconductor quantum dots. *Nat. Biotechnol.* 2002, *21*, 41-46.

[24] Dai, Q.; Li, D.; Chang, J.; Song, Y.; Kan, S.; Chen, H.; Zou, B.; Xu, W.; Xu, S.; Liu, B.; et al. Facile synthesis of magic-sized CdSe and CdTe nanocrystals with tunable existence periods. *Nanotechnology* 2007, *18*, 405603, doi:10.1088/0957-4484/18/40/405603.

[25] Murray, C.B.; Kagan, C.R.; Bawendi, M.G. Synthesis and Characterization of Monodisperse Nanocrystals and Close-Packed Nanocrystal Assemblies. *Annu. Rev.Mater* 2000, *30*, 545.

[26] Chen, X.; Samia, A.C.S.; Lou, Y.; Burda, C. Investigation of the crystallization process in 2 nm CdSe quantum dots. *J. Am. Chem. Soc.* 2005, *127*, 4372-4375, doi:10.1021/ja0458219.

[27] Riehle, F.S.; Bienert, R.; Thomann, R.; Urban, G.A. Blue Luminescence and Superstructures from Magic Size Clusters of CdSe. *Nano Lett.* 2009, *9*, 514-518, doi:10.1021/nl080150o.

[28] McBride, J.R.; Dukes, A.D.; Schreuder, M. a; Rosenthal, S.J. On Ultrasmall Nanocrystals. *Chem. Phys. Lett.* 2010, *498*, 1-9, doi:10.1016/j.cplett.2010.08.052.

[29] Dukes, A.D.; McBride, J.R.; Rosenthal, S.J. Synthesis of Magic-Sized CdSe and CdTe Nanocrystals with Diisooctylphosphinic Acid. *Chem. Mater.* 2010, *22*, 6402-6408, doi:10.1021/cm102370a.

[30] Bowers, M.J.; McBride, J.R.; Rosenthal, S.J. White-light emission from magic-sized cadmium selenide nanocrystals. *J Am Chem Soc* 2005, *127*, 15378-15379.

[31] Harrell, S.M.; McBride, J.R.; Rosenthal, S.J. Synthesis of ultrasmall and magic-sized CdSe nanocrystals. *Chem. Mater.* 2013, *25*, 1199-1210.

[32] Nguyen, K.A.; Day, P.N.; Pachter, R. Understanding structural and optical properties of nanoscale CdSe magic-size quantum dots: Insight from computational prediction. *J. Phys. Chem. C* 2010, *114*, 16197-16209, doi:10.1021/jp103763d.

[33] Shashkova, S.; Leake, M.C. Single-molecule fluorescence microscopy review: Shedding new light on old problems. *Biosci. Rep.* 2017, *37*, BSR20170031, doi:10.1042/BSR20170031.

[34] Martynov, V.I.; Pakhomov, A.A.; Popova, N. V.; Deyev, I.E.; Petrenko, A.G. Synthetic Fluorophores for Visualizing Biomolecules in Living Systems. *Acta Naturae* 2016, *8*, 33-46, doi:10.32607/20758251-2016-8-4-33-46.

[35] Yan, F.; Fan, K.; Bai, Z.; Zhang, R.; Zu, F.; Xu, J.; Li, X. Fluorescein applications as fluorescent probes for the detection of analytes. *TrAC - Trends Anal. Chem.* 2017, *97*, 15-35, doi:10.1016/j.trac.2017.08.013.

[36] Almeida Silva, A.; Silva, M.J.; da Luz, F.A.; Silva, D.; de Deus, S.; Dantas, N. Controlling the Cytotoxicity of CdSe Magic-Sized Quantum Dots as a Function of Surface Defect Density. *Nano Lett. 14*, 5452-5457, doi:10.1021/nl5028028.

[37] Silva, A.C.A.; Deus, S.L.V. De; Silva, M.J.B.; Dantas, N.O. Highly stable luminescence of CdSe magic-sized quantum dots in HeLa cells. *Sensors Actuators, B Chem.* 2014, *191*, 108-114, doi:10.1016/j.snb.2013.09.063.

[38] Silva, A.C.A.; Dantas, N.O.; Silva, M.J.B.; Spanó, A.M.; Goulart, ; Luiz Ricardo Functional Nanocrystals : Towards Biocompatibility , Nontoxicity and. In *Advances in Biochemistry & Applications in Medicine*; 2017; pp. 1-27.

[39] Dias, E.H.V.; Pereira, D.F.C.; de Sousa, B.B.; Matias, M.S.; de Queiroz, M.R.; Santiago, F.M.; Silva, A.C.A.; Dantas, N.O.; Santos-Filho, N.A.; de Oliveira, F. In vitro tracking of phospholipase A 2 from snake venom conjugated with magic-sized quantum dots. *Int. J. Biol. Macromol.* 2019, *122*, 461-468, doi:10.1016/j.ijbiomac.2018.10.185.

[40] Silva, A.C.A.; Correia, L.I.V.; Silva, M.J.B.; Zóia, M.A.P.; Azevedo, F.V.P.V.; Rodrigues, Jéssica Peixoto Goulart, L.R.; Ávila, Veridiana de Melo Dantas, N.O. Biocompatible Magic Sized Quantum Dots: Luminescent Markers and Probes. In; Correia, L.I.V., Ed.; IntechOpen: Rijeka, 2018; p. Ch. 6 ISBN 978-1-78923-295-0.

[41] Silva, A.C.A.; Azevedo, F.V.P.V.; Zóia, M.A.P.; Rodrigues, J.P.; Dantas, N.O.; Melo, V.R.Á.; Goulart, L.R. Magic Sized Quantum Dots as a Theranostic Tool for Breast Cancer. In *Recent Studies & Advances in Breast Cancer*; Open Access eBooks: Wilmington, 2017; pp. 1-10 ISBN 978-81-935757-2-7.

[42] Silva, A.C.A.; Neto, E.S.F.; Da Silva, S.W.; Morais, P.C.; Dantas, N.O. Modified phonon confinement model and its application to CdSe/CdS core-shell magic-sized quantum dots synthesized in aqueous solution by a new route. *J. Phys. Chem. C* 2013, doi:10.1021/jp308500r.

[43] de França, C.C.L.; Meneses, D.; Silva, A.C.A.; Dantas, N.O.; de Abreu, F.C.; Petroni, J.M.; Lucca, B.G. Development of novel paper-based electrochemical device modified with CdSe/CdS magic-sized quantum dots and application for the sensing of dopamine. *Electrochim. Acta* 2021, *367*, 137486, doi:10.1016/j.electacta.2020.137486.

[44] de Lima França, C.C.; da Silva Terto, E.G.; Dias-Vermelho, M. V.; Silva, A.C.A.; Dantas, N.O.; de Abreu, F.C. The electrochemical behavior of core-shell CdSe/CdS magic-sized

quantum dots linked to cyclodextrin for studies of the encapsulation of bioactive compounds. *J. Solid State Electrochem.* 2016, *20*, 2533-2540, doi:10.1007/s10008-016-3221-8.

[45] Dooley, C.T.; Dore, T.M.; Hanson, G.T.; Jackson, W.C.; Remington, S.J.; Tsien, R.Y. Imaging dynamic redox changes in mammalian cells with green fluorescent protein indicators. *J. Biol. Chem.* 2004, *279*, 22284-22293, doi:10.1074/jbc.M312847200.

[46] Gutscher, M.; Pauleau, A.L.; Marty, L.; Brach, T.; Wabnitz, G.H.; Samstag, Y.; Meyer, A.J.; Dick, T.P. Real-time imaging of the intracellular glutathione redox potential. *Nat. Methods* 2008, *5*, 553-559, doi:10.1038/nmeth.1212.

[47] Hanson, G.T.; Aggeler, R.; Oglesbee, D.; Cannon, M.; Capaldi, R.A.; Tsien, R.Y.; Remington, S.J. Investigating Mitochondrial Redox Potential with Redox-sensitive Green Fluorescent Protein Indicators. *J. Biol. Chem.* 2004, *279*, 13044-13053, doi:10.1074/jbc.M312846200.

[48] Silva, A.C.A.; Correia, L.I.V.; Silva, M.J.B.; Zóia, M.A.P.; Azevedo, F.V.P.V.; Rodrigues, J.P.; Goulart, L.R.; Ávila, Veridiana de Melo Dantas, N.O. Biocompatible Magic Sized Quantum Dots: Luminescent Markers and Probes. In *State of the Art in Nano-Bioimaging*; 2017; Vol. 1, pp. 95-104 ISBN 9789537619992.

[49] Silva, A.C.A.; Azevedo, F.V.P.V.; Zóia, M.A.P.; Rodrigues, J.P.; Dantas, N.O.; Melo, V.R.Á.; Goulart, L.R. Magic Sized Quantum Dots as a Theranostic Tool for Breast Cancer. In *Recent Studies & Advances in Breast Cancer*; 2017 ISBN 978-81-935757-2-7.

[50] Silva, A.C.A.; Freschi, A.P.P.; Rodrigues, C.M.; Matias, B.F.; Maia, L.P.; Goulart, L.R.; Dantas, N.O. Biological analysis and imaging applications of CdSe/CdSxSe1−x/CdS core–shell magic-sized quantum dot. *Nanomedicine Nanotechnology, Biol. Med.* 2016, *12*, 1421-1430, doi:10.1016/j.nano.2016.01.001.

[51] Vlasceanu, G.; Grumezescu, A.M.; Gheorghe, I.; Chifiriuc, M.C.; Holban, A.M. Quantum dots for bioimaging and therapeutic applications. In *Nanostructures for Novel Therapy: Synthesis, Characterization and Applications*; Elsevier Inc., 2017; pp. 497-515 ISBN 9780323461481.

[52] Li, J.; Wu, D.; Miao, Z.; Zhang, Y. Preparation of Quantum Dot Bioconjugates and their Applications in Bio-Imaging. *Curr. Pharm. Biotechnol.* 2010, *11*, 662-671, doi:10.2174/138920110792246582.

[53] Gondim, B.L.C.; da Silva Catarino, J.; de Sousa, M.A.D.; de Oliveira Silva, M.; Lemes, M.R.; de Carvalho-Costa, T.M.; de Lima Nascimento, T.R.; Machado, J.R.; Rodrigues, V.; Oliveira, C.J.F.; et al. Nanoparticle-Mediated Drug Delivery: Blood-Brain Barrier as the Main Obstacle to Treating Infectious Diseases in CNS. *Curr. Pharm. Des.* 2019, *25*, 3983-3996, doi:10.2174/1381612825666191014171354.

[54] Makride, S.C.; Gasbarro, C.; Bello, J.M. Bioconjugation of quantum dot luminescent probes for Western blot analysis. *Biotechniques* 2005, *39*, 501-506, doi:10.2144/000112004.

[55] Field, L.D.; Walper, S.A.; Susumu, K.; Lasarte-Aragones, G.; Oh, E.; Medintz, I.L.; Delehanty, J.B. A Quantum Dot-Protein Bioconjugate That Provides for Extracellular Control of Intracellular Drug Release. *Bioconjug. Chem.* 2018, *29*, 2455-2467, doi:10.1021/acs.bioconjchem.8b00357.

[56] Kwon, J.; Jun, S.W.; Choi, S.I.; Mao, X.; Kim, J.; Koh, E.K.; Kim, Y.H.; Kim, S.K.; Hwang, D.Y.; Kim, C.S.; et al. FeSe quantum dots for in vivo multiphoton biomedical imaging. *Sci. Adv.* 2019, *5*, doi:10.1126/sciadv.aay0044.

Chapter 2

Design of Coelenterazine Analogue to Reveal Bioluminescent Reaction of Human Serum Albumin

Ryo Nishihara, Kazuki Niwa, Tatsunosuke Tomita and Ryoji Kurita

Abstract

This chapter describes the design of an imidazopyrazinone-type luciferin named as HuLumino**1** by us and investigation of its luminescence properties. This luciferin was designed to generate bioluminescence by human serum albumin (HSA) rather than by luciferase derived from luminous organisms. HuLumino**1** was developed by modifying a methoxy-terminated alkyl chain to the C-6 position and eliminating a benzyl group at the C-8 position of coelenterazine. To clarify the basis of light emission by HSA, the detailed kinetic properties of the HuLumino**1**/HSA pair were investigated using a calibrated luminometer. The enzymatic oxidation of HuLumino**1** was observed only in the presence of HSA. Results of HSA quantification experiments using HuLumino**1** agreed with less than 5% differences with those of enzyme-linked immunosorbent assays, suggesting HuLumino**1** could be used for quantitative analysis of HSA levels in serum samples without any pretreatments. These results demonstrate the advantages of the coelenterazine analogue as a bioluminescence reagent to detect non-labeled proteins, which generally do not function as enzymes.

Keywords: Bioluminescence, Coelenterazine, Luciferin, Luciferase, Human serum albumin, Quantum yield, Enzyme-linked immunosorbent assay

1. Introduction

Luminous organisms, such as *fireflies*, *Gaussia princeps*, *Oplophorus gracilirostris*, and *sea pansy Renilla reniformis*, generate bioluminescence (BL), which is the light emission in the absence of external energy sources [1, 2]. In the core of BL, an enzymatic reaction occurs involving a bioluminescent substrate (luciferin) and enzyme (luciferase). In general, the enzymatic luminescence reaction proceeds between a specific luciferin-luciferase pair to allow for highly sensitive and specific detection/imaging of diverse molecular events in living subjects [2–4].

However, in some cases, bioluminescent or chemiluminescent substrates may induce enzymatic luminescence activity of non-bioluminescence enzymes. For example, CycLuc2, a synthetic analog of firefly luciferin, can be catalyzed to emit light by long-chain fatty acid acyl-CoA synthetase found in non-luminous insects

[5–7]. In addition, the heme-containing enzyme myeloperoxidase, which is abundantly expressed in neutrophils and monocytes, can catalyze the luminescence reaction of xenobiotic luminol [8, 9]. These reports suggested that the introduction of appropriate exogenous luminescent substrates reveals luminous activity of non-bioluminescence enzymes, which can significantly differ from the conventional function of the enzyme and has potential for use in quantitative analysis of enzymes without any labeling procedures, including transgene introduction of luciferase from luminous organisms. Here, we describe the design and bioluminescence characterization of a luciferin analogue which was selectively catalyzed to exhibit bioluminescence by human serum albumin (HSA) [10]. Serum albumins perform various physiological functions; they maintain colloid osmotic blood pressure and transport several exogenous and endogenous molecules. However, they are not categorized in the list of EC number, indicating they are not considered typical enzymes. The bioluminescence system of HSA with the luciferin analogue synthesized by us is novel and different from the conventional luciferin-luciferase reaction systems.

2. Design of coelenterazine analogue

2.1 Coelenterazine analogue with HSA-specific bioluminescence

Most luciferases from luminous marine organisms use coelenterazine (CTZ) as their luciferin to form coelenteramide in an excited state, with emission ranging from blue to green at approximately 400–500 nm (**Figure 1**) [11, 12]. CTZ is oxidized by bovine serum albumin (BSA) in addition to luciferase, and this has been considered as nonspecific reaction mainly occurs because of a simple luminescence reaction that requires only an oxygen molecule [1] (**Figure 1**).

The emission ability of CTZ is derived from the imidazopyrazinone ring, and the chemical structure of sidechains at the C-2, C-6, and C-8 position of the imidazopyrazinone core significantly affect enzyme recognition. For example, *Cypridina* luciferase oxidizes only *Cypridina* luciferin, which contains a basic guanidine moiety at the C-8 position of the imidazopyrazinone ring, not CTZ (**Figure 2**) [13]. A mutant *Oplophorus* luciferase (NanoLuc) uses furimazine rather than CTZ, and is known as a versatile reporter of BL (**Figure 2**) [14]. Thus, each bioluminescence probe has been individually developed with an imidazopyrazinone analogue that is suitable for the geometry of the active site in the pocket of mutant luciferase. We also reported that RLuc8.6-535SG, a mutant *R. reniformis* luciferase, utilizes BottleBlue2.3 (BBlue2.3), a CTZA that can permeate the cell membrane and emits bright visible luminescence suitable for deep-tissue imaging of cancer cells in vivo (**Figure 3a**) [15].

To clarify the potential enzymatic luminescence activity of human proteins, we focused on HSA, which accounts for approximately 65% of serum proteins in the human body [16]. This abundant protein is involved in a wide variety of

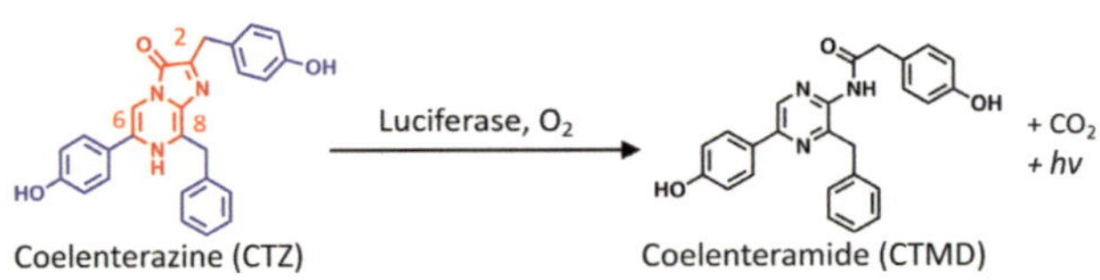

Figure 1.
Chemical reaction of coelenterazine (CTZ)-dependent bioluminescence. The imidazopyrazinone structure and modifiable substituent are highlighted in red and blue, respectively.

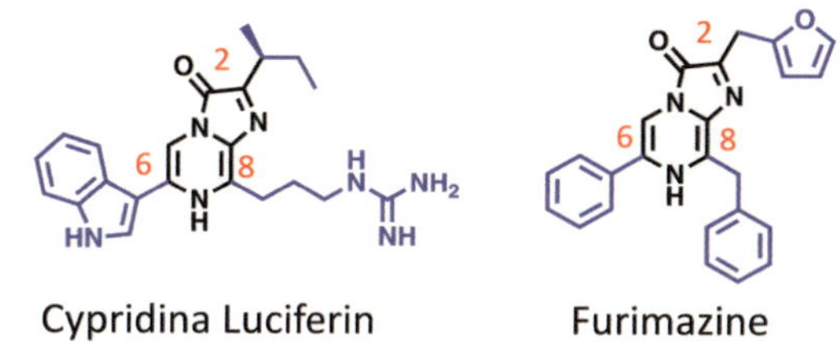

Figure 2.
Chemical structures of imidazopyrazinone-based analogues.

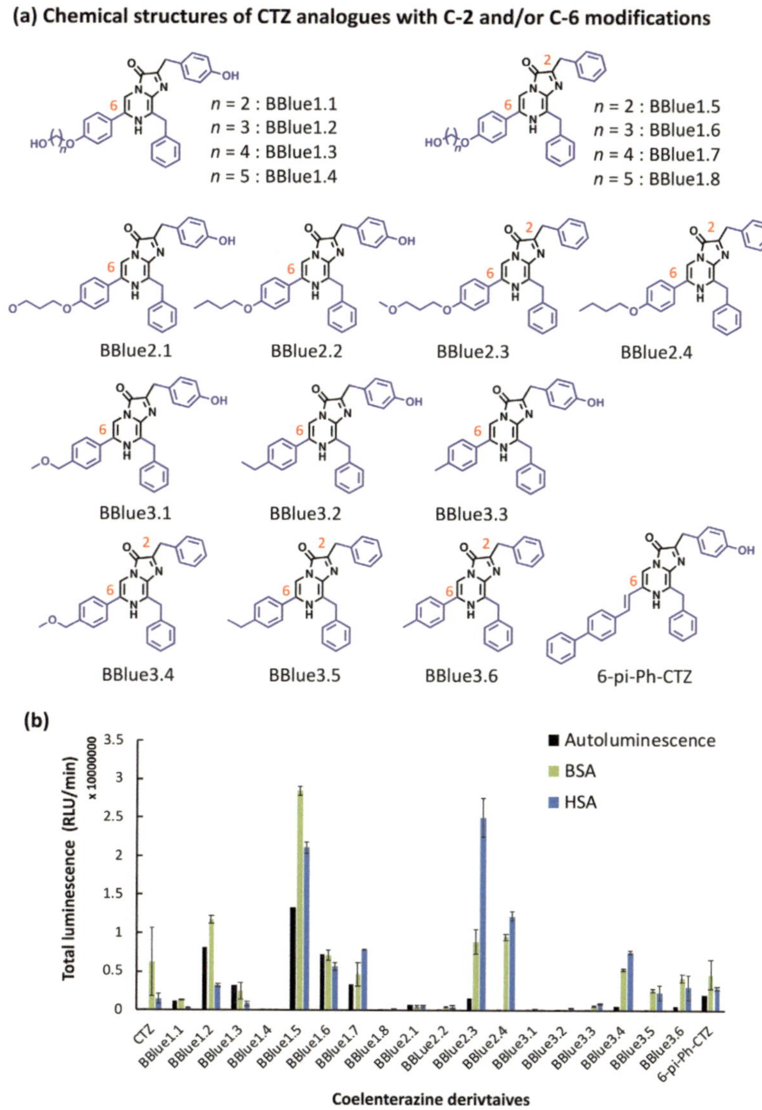

Figure 3.
Chemical structures of coelenterazine analogues (CTZAs) upon (a) C-2 and/or C-6 substitution. (b) Luminescence intensities obtained with serum albumin (0.1 mg/mL).

physiological functions, such as maintaining osmotic pressure, buffering blood pH levels, and carrying ligands including hormones, amino acids, and fatty acids [17, 18]. In addition, HSA and some ligand complexes often possess enzymatic activities, such as Kemp elimination and hydrolysis of esters because of their unique ability to bind small hydrophobic molecules in some cavities; however, the potential enzymatic activities remain unclear [17, 19].

First, to obtain a rational luciferin with an imidazopyrazinone core for HSA-specific BL, we assayed CTZ and previously reported 18 CTZAs named as Bottle Blue (BBlue), where the *p*-hydroxy phenyl group at the C-6 position of CTZ was modified by alkylation (**Figure 3**), with serum albumins (fatty acid free HSA and BSA). In this chapter, except for in Section 2.2, luminescence measurements were performed using luminometers (GloMax20/20 Luminometer or GloMax Explorer Multimode Microplate Reader) manufactured by Promega (Madison, WI, USA). Next, BBlue2.3, a CTZA with a methoxy-terminated alkyl linker chain of three methylene units at the C-6 position, exhibited the brightest emission, which produced 16.6-fold higher luminescence when combined with HSA (i.e. BBlue2.3/HSA pair) compared to that of the CTZ/HSA pair (**Figure 3**).

Based on these results, we predicted that elimination of the benzyl group at the C-8 position of BBlue2.3 would relieve its steric hindrance with key amino acids in the substrate binding site of HSA and enhance the enzymatic luminescence reaction of HSA. We then designed and synthesized a novel CTZA, named as Human Luminophore **1** (HuLumino**1**) based on the synthetic procedures of BBlue2.3 [15] and an array of 5 CTZAs containing known analogues [20] to investigate the effect of substitution at the C-2, C-6, and C-8 positions of CTZ on serum albumin-dependent luminescence (**Figure 4b–d**). Moreover, the luminescence of

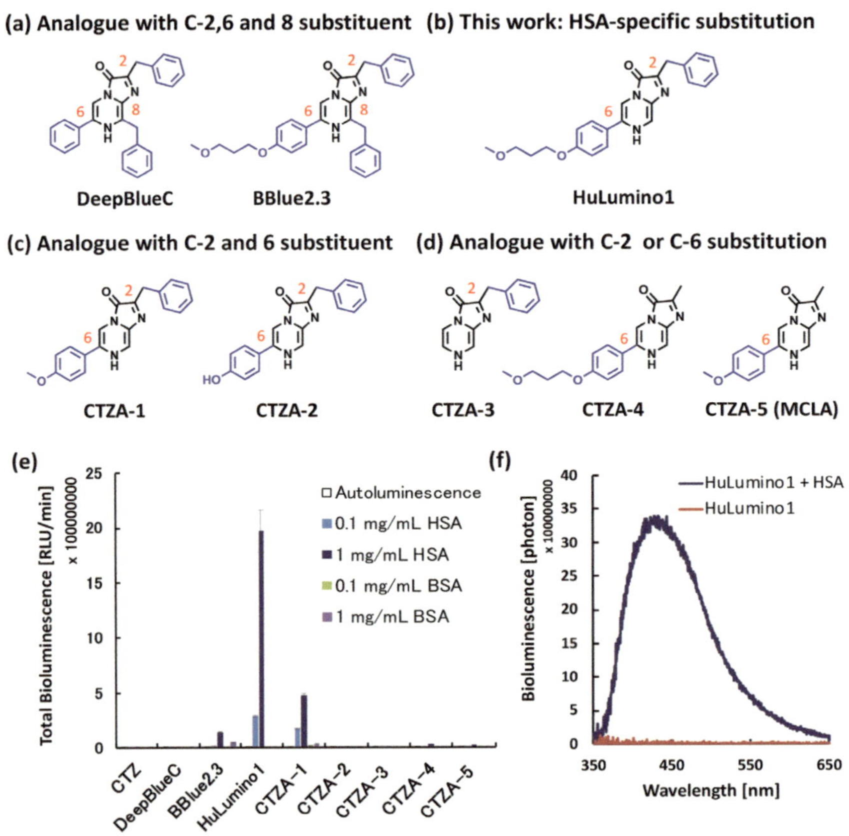

Figure 4.
Chemical structures of coelenterazine analogues (CTZAs) upon (a) C-2, C-6, and C-8 substitutions, (b) HSA-specific substitution, (c) C-2 and C-6 substitution, and (d) C-2 or C-6 substitution. (e) Luminescence from serum albumins (0.1 or 1 mg/mL) treated with the indicated substrate (10 μM); error bars represent the standard deviations of three measurements. (f) Bioluminescence spectra of HuLumino1 in the presence or absence of HSA.

commercially available CTZAs (DeepBlueC and MCLA) was compared with the synthesized CTZAs when fatty acid free HSA and BSA were added (**Figure 4e**). Although the autoluminescence levels in PB buffer of HuLumino**1** was similar to that of CTZ, only HuLumino**1** displayed significantly enhanced luminescence dependent on HSA (but not BSA containing fatty acid) concentrations, as indicated by the 14.1-fold higher emission compared with that of the BBlue2.3/HSA pair. Furthermore, the luminescence intensity of the HuLumino**1**/HSA pair was found to be 718-fold higher than that of the CTZ/BSA pair, which has been reported to produce luminescence [11, 12]. The HuLumino**1**/HSA pair exhibited flash-type luminescence with a peak wavelength of 432 nm. (**Figure 4f**). Unexpectedly, HuLumino**1** selectively activated HSA by recognizing the subtle conformational difference in the substrate binding site, although the overall sequence homology between HSA and BSA is 75.6%.

These results suggest "luciferase" activity of HSA catalyzes the enzymatic luminescence reaction of CTZAs to produce "bioluminescence".

2.2 Quantitative evaluation of luminescence intensity

To characterize the enzymatic luminescence reaction with HSA, the bioluminescence intensity of CTZAs and HSA pairs was quantitatively evaluated. Bioluminescence intensity is generally determined by several reaction factors including the bioluminescence quantum yield (φ_{BL}) of luciferin, turnover number (k_{cat}), and active luciferase concentration. Kinetic parameters were determined using a custom-built luminometer with a photomultiplier tube (PMT) (H11890–01; Hamamatsu Photonics, Japan), and its absolute responsibility for total number of emitted photons in the instrument was calibrated for luminescence spectrum of each luciferin-luciferase pair [21, 22]. The absolute responsibility of the luminometer for the CTZ-utilizing bioluminescent system was determined as described previously [21]. The φ_{BL} values were calculated from the total number of emitted photons and total number of reacted luciferin molecules. To integrate all photons derived from the enzymatic reaction, the number of photons was monitored using the luminometer from before initiating the reaction to until the reaction was completed. The reaction was initiating by injection of fatty acid-free HSA PB solution (100 μg/mL or 10 mg/mL) into the preinstalled luciferin PB solution (20 nM) in the luminometer.

The Michaelis–Menten constant (K_m) of luciferin was calculated from Lineweaver-Burk plots constructed using a standard method. The catalytic constant (k_{cat}), which is the turnover number of the reaction for luciferin by a single luciferase molecule per second, was calculated from the φ_{BL} value and maximum velocity (V_{max}) determined from the Lineweaver-Burk plots.

The apparent K_m of the HuLumino**1**/HSA pair was 4.3 μM, which was comparable to that of the NanoLuc system [23]. In contrast, the k_{cat} value of the NanoLuc system was 294-fold higher than that of the HuLumino**1**/HSA pair, and the catalytic efficiency of the luminescence reaction by HSA was lower than that of conventional luciferase [24]. However, in the luminescence system with HSA, HuLumino**1** displayed a slightly higher K_m value than CTZA-1, but its k_{cat} value was approximately 3-fold higher (**Table 1**), resulting in 4.2-fold stronger light emission than that of CTZA-1 (**Figure 4e**). For luciferin with high enzyme affinity (e.g., CTZA −1), oxyluciferin, a product of the BL reaction, appears to competitively inhibit the luminescence reaction [25]. Moreover, HuLumino**1** showed an enzyme affinity and bioluminescence quantum yield (φ_{BL}) of more than 5- and 100-fold higher than those of CTZ, respectively. The detailed luminescent profiles suggest that the structural properties of the alkyl linker chain modified at the C-6 position of the

Pair	$\varphi_{BL}{}^{a}[\times 10^{-5}]$	$K_m{}^{b}[\mu M]$	$k_{cat}{}^{c}[s^{-1}]$
CTZ/HSA	0.32 ± 0.03	25.3 ± 5.2	2.75 ± 0.3
HuLumino1/HSA	30.9 ± 3.1	4.28 ± 1.24	0.30 ± 0.06
CTZA-4/HSA	42.2 ± 9.1	2.46 ± 0.40	0.11 ± 0.09

[a–c]*Errors represent standard error of the mean values for triplicate experiments.*
[b]*Michaelis-Menten constant (K_m) values were determined by Lineweaver-Burk plots via measurements of initial rates of light emission over a range 0.5 to 20 μM.*
[c]*Turnover rate (k_{cat}) values were calculated by dividing maximum velocity (V_{max}) by the φ_{BL}. The V_{max} were determined by Lineweaver-Burk plots.*

Table 1.
Luminescent profiles of CTZ and its analogues with HSA.

imidazopyrazinone core contribute to the efficient HSA-catalyzed emission reaction. Although the catalytic efficiency of HuLumino**1**/HSA is much lower than that of the NanoLuc system, HuLumino**1** is a luciferin that is relatively more suitable for HSA than other existing luminescent substrates.

2.3 Enzymatic reaction site of HSA

The crystal structure of HSA, with binding to a variety of drugs, clarified the two principal drug binding sites in different subdomains (site 1 in subdomain IIA and site 2 in subdomain IIIA) [26, 27]. To investigate the luminescent reaction site between HSA and HuLumino**1**, a competitive assay was conducted with two site-specific HSA drugs (warfarin-site1 and ibuprofen-site2) [27]. Fatty acid free HSA PB solution was pre-treated with the drugs (0–100 μM) to fill binding site 1 or 2 before adding HuLumino**1**. The luminescence of HuLumino**1** was negligibly low in the presence of HSA-ibuprofen complex (**Figure 5a**). In contrast, HuLumino**1** exhibited efficient luminescence with the HSA-warfarin complex, indicating that HuLumino**1** selectivity binds to the site 2 cavity of HSA. In detailed analysis of the inhibitory kinetics, Lineweaver-Burk plots displayed that ibuprofen competitively inhibited the binding of HuLumino**1** to HSA (K_i of ibuprofen was 6.3 nM) (**Figure 5b**). Therefore, the enzymatic reaction site of HuLumino**1** was experimentally determined to be binding site 2. Next, docking simulation with the Molecular Operating Environment software package was carried out to predict the binding poses. The simulation displayed the specific binding of HuLumino**1** to the hydrophobic cavity of site 2 by interacting with several amino acids including R410, K414, and L453 (**Figure 5c–e**). Particularly, R410, a key amino acid residue in the esterase activity of HSA [17], is also involved in the luminescence reaction (**Figure 5e**).

Next, to investigate the effect of the steric structure of HSA on luminescence, HSA pretreated with 10 M guanidine hydrochloride, a reagent commonly used to induce denaturation of the α-helix structure of proteins [28], was prepared, and the luminescence of HuLumino**1** was extremely low in the presence of denatured HSA (data not shown). Therefore, the enzymatic reaction of HuLumino **1** depends on the microenvironment and steric structure such as binding site 2 constructed by the folding structure of HSA. These results suggest that emission of HuLumino**1**/HSA is not a non-specific chemiluminescence commonly found in other imidazopyrazinone compounds.

2.4 Bioluminescent assay for HSA

Low levels of HSA in the serum (<35 mg/mL) are biomarkers of several diseases such as malnutrition, cirrhosis, and chronic hepatitis [29]. In hospitals, HSA is

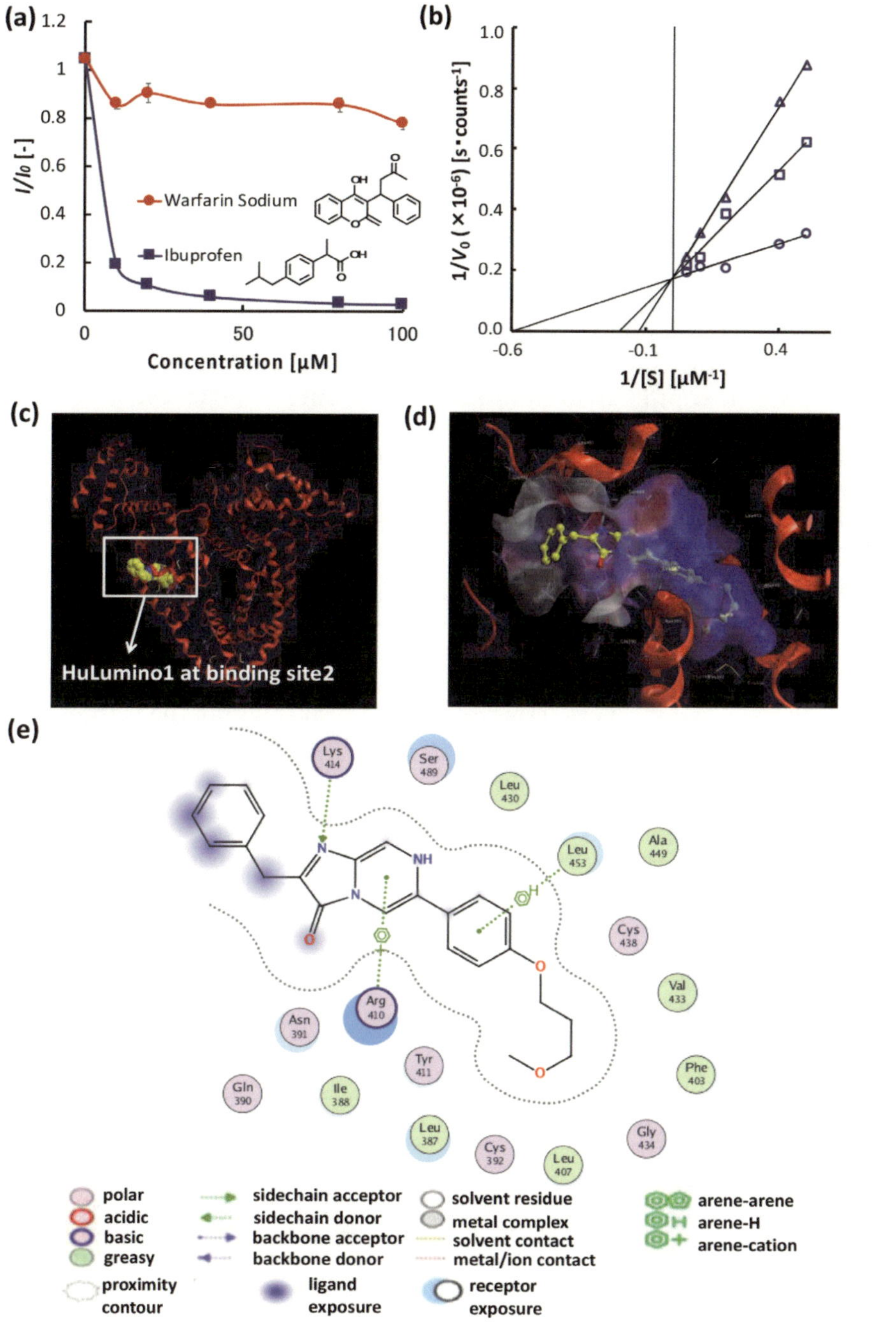

Figure 5.
(a) Luminescence response in the presence (I) vs. absence (I_0) of the binding drug concentration (0–100 μM). (b) Lineweaver-Burk plot indicating competitive inhibition by ibuprofen. V_0 is defined as the luminescence intensity over the initial 30 s and [S] is the substrate concentration. The concentrations of ibuprofen were 1 nM (square) and 5 nM (triangle). Open circles indicate negative controls with no ibuprofen. (c) Ligand-binding site 2 of HSA with HuLumino1 as posed by the molecular operating environment software (d) magnified view of binding site 2. The ligand binding site in the blue region indicates the presence of the hydrophobic environment. (e) Predicted interaction between HuLumino1 and HSA.

evaluated using the colorimetric bromocresol green assay or ELISA. Both can provide a reliable assessment of albumin but require sample preparation and processing time (e.g. 3 h for ELISA) [30]. Therefore, an assay for the simple, accurate and rapid detection of HSA in the serum should be developed for clinical diagnosis. We demonstrated that the BL assay can be used to evaluate HSA based on the enzymatic luminescence reaction of HuLumino**1**. Regarding the selectivity of the reaction of HuLumino**1**, no reactivity with other proteins (BSA, β-galactosidase,β-lactoglobulin, catalase, α-chymotrypsinogen, hemoglobin, human IgG, porcine lipase, papain, pepsin, trypsin, γ-globulin, carbonic anhydrase, concanavalin A, glucosidase, histone, myoglobin, and RNase, 0.1 mg/mL) was observed, and only HSA led to distinct luminescence enhancement (**Figure 6a**). Although the coexistence of most proteins did not affect the enzymatic reaction of the HSA/Hulumino**1** pair, a slight decrease in luminescence was observed in the presence of some proteins (data not shown). This indicates that HuLumino**1** nonspecifically binds to other proteins but does not exhibit BL. Hence, HuLumino**1** can be used to detect HSA without interference from other proteins, as it exhibited excellent selectivity for HSA even in a complicated biological system. The luminescence of HuLumino**1** was enhanced in an HSA concentration-dependent manner and exhibited a constant intensity at HSA concentrations above 10 mg/mL (**Figure 6b**). A linear increase in luminescence, within the spiked HSA concentration range of 0–0.1 mg/mL in PBS-diluted serum, resulted in a detection limit of 8.6 µg/mL for HSA, which was comparable to the standard detection limit of HSA in physiological systems (**Figure 6c**) [31].

Finally, two HSA assays, including our developed BL-based assay and ELISA, were performed to evaluate human serum from male AB plasma. The HSA levels calculated with HuLumino**1** agreed well with those estimated by ELISA within 5% error. The spike and recovery tests also showed results within the margin of 7%

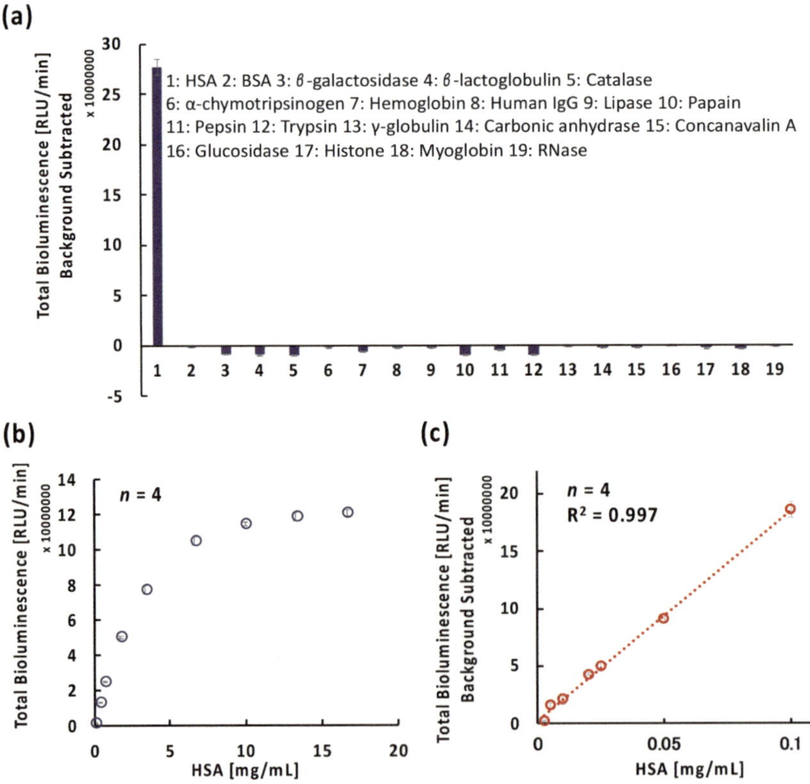

Figure 6.
(a) Variation in luminescence of HuLumino1 (10 µM) in the presence of proteins (0.1 mg/mL); error bars represent the standard deviations of three measurements. (b) Luminescence intensity of HuLumino1 (20 µM) containing various concentrations of HSA (0–17 mg/mL) in PBS and (c) HSA (0–17 mg/mL) in PBS-diluted plasma (100-fold, pH 7.4).

Amount of HSA added (mg/mL)	HSA (mg/mL) determined by developed method[a]	HSA (mg/mL) determined by ELISA	Recovery
0	39.0 ± 3.1	41.0 ± 3.6	95.2
1	44.5 ± 0.5	ND	106.1
2.5	45.2 ± 0.5	ND	104.1

ND: Not Determined.
[a]*Conditions: HuLumino**1** (20 μM) in PBS-diluted serum (1000-fold, 10 mM, pH 7.4).*

Table 2.
Assay of HSA in human serum.

error (**Table 2**). Therefore, the BL-based HSA assay showed analytical capability with high sensitivity and could detect HSA within 10 min including the sample preparation and measurement times. In addition, we detected the expression of recombinant HSA in living COS-1 cells (data not shown), indicating that HuLumino**1** can be used in molecular biology studies and in biomedical applications.

We designed and synthesized the first luciferin (HuLumino**1**), an analogue with C-6 and C-8 modification of CTZ, which exhibited bioluminescence with HSA. HuLumino**1** rapidly detected HSA with high sensitivity and specificity, even in real human plasma containing various interfering biomolecules. Detailed kinetic investigation of the enzymatic reaction clarified the enzyme recognition of HuLumino**1** from HSA drug binding site 2, resulting a highly selective reaction and revealing a reaction with both native HSA and recombinant HSA expressed in COS-1 cells. Therefore, the BL-based assay with HuLumino**1**, either used alone or coupled with ELISA, can be used for the early diagnosis of HSA-related diseases, enabling accurate and rapid detection of HSA in serum samples without pre-treatment. The information obtained through detailed investigation of the HuLumino**1**/HSA pair may be extended to protein assays based on a luminescent reaction without genetically engineered luciferases.

Acknowledgements

We acknowledge funding from JST, PRESTO Grant Number JPMJPR20EB, JSPS KAKENHI Grant Number 20 K15421, AMED under Grant Number JP191m0203012, Shimazu Science Foundation, Research Foundation for Opto-Science and Technology, and Nakatani Foundation for Advancement of Measuring Technologies in Biomedical Engineering, and DAICENTER project grant from DBT (Govt. of India) to Renu Wadhwa and special strategic grant from AIST (Japan).

Conflict of interest

The authors declare no conflict of interest.

Notes/thanks/other declarations

We thank Dr. Tsukasa Ishihara (Biomedical Research Institute, National Institute of Advanced Industrial International Science and Technology) for his help with the docking simulation experiment.

Author details

Ryo Nishihara[1*], Kazuki Niwa[1], Tatsunosuke Tomita[1] and Ryoji Kurita[1,2]

1 National Institute of Advanced Industrial Science and Technology (AIST), Tsukuba, Japan

2 University of Tsukuba, Tsukuba, Japan

*Address all correspondence to: r.nishihara@aist.go.jp

References

[1] Shimomura O. Bioluminescence: Chemical Principles and Methods. Bioluminescence: Chemical Principles and Methods. Singapore: World Scientific Publishing Company; 2006. 471 p.

[2] Badr C. Bioluminescent Imaging: Methods and Protocols. Bioluminescent Imaging: Methods and Protocols. New York: Springer; 2014. 1276 p.

[3] Rowe L, Dikici E, Daunert S. Engineering bioluminescent proteins: expanding their analytical potential. Analytical Chemistry. 2009;81(21): 8662–8668. DOI: 10.1021/ac9007286

[4] Ozawa T, Yoshimura H, Kim S. Advances in fluorescence and bioluminescence imaging. Analytical Chemistry. 2013;85(2):590–609. DOI: 10.1021/ac3031724

[5] Mofford D, Reddy G, Miller S. Latent luciferase activity in the fruit fly revealed by a synthetic luciferin. Proceedings of the National Academy of Sciences of the United States of America. 2014;111(12):4443–4448. DOI: 10.1073/pnas.131900111

[6] Mofford D, Liebmann K, Sankaran G, Reddy G, Reddy G, Miller S. Luciferase activity of insect fatty acyl-CoA synthetases with synthetic luciferins. ACS Chemical Biology. 2017;12(12):2946–2951. DOI: 10.1021/acdchembio.7b00813

[7] Mofford D, Adams S, Reddy G, Reddy G, Miller S. Luciferin amides enable in vivo bioluminescence detection of endogenous fatty acid amide hydrolase activity. Journal of the American Chemical Society. 2015;137 (27):8684–8687. DOI: 10.1021/ jacs.5b04357

[8] Liu R, Tang J, Xu Y, Dai Z. Bioluminescence imaging of inflammation in vivo based on bioluminescence and fluorescence resonance energy transfer using nanobubble ultrasound contrast agent. ACS Nano. 2019;13(5):5124–5132. DOI: 10.1021/acsnano.8b08359.

[9] Gross S, Gammon S, Moss B, Rauch D, Harding J, Heinecke J, et al. Bioluminescence imaging of myeloperoxidase activity in vivo. Nature Medicine. 2009;15(4):455–461. DOI: 10.1038/nm.1886

[10] Nishihara R, Niwa K, Tomita T, Kurita R. Coelenterazine analogue with human serum albumin-specific bioluminescence. Bioconjugate Chemistry. 2020;31(12):2679–2684. DOI: 10.1021/acs.bioconjchem. 0c00536

[11] Vassel N, Cox C, Naseem R, Morse V, Evans R, Power R, et al. Enzymatic activity of albumin shown by coelenterazine chemiluminescence. Luminescence. 2012;27(3):234–241. DOI: 10.1002/bio.2357

[12] Lindberg E, Mizukami S, Ibata K, Miyawaki A, Kikuchi K. Development of luminescent coelenterazine derivatives activatable by beta-galactosidase for monitoring dual gene expression. Chemistry-A European Journal. 2013;19(44):14970–14976. DOI: 10.1002/chem.201302002

[13] Maguire C, Bovenberg M, Crommentuijn M, Niers J, Kerami M, Teng J, et al. Triple bioluminescence imaging for in vivo monitoring of cellular processes. Molecular Therapy-Nucleic Acids. 2013;2(6):e99. DOI: 10.1038/mtna.2013.25

[14] Hall M, Unch J, Binkowski B, Valley M, Butler B, Wood M, et al. Engineered luciferase reporter from a deep sea shrimp utilizing a novel imidazopyrazinone substrate. ACS

Chemical Biology. 2012;7(11):1848–1857. DOI: 10.1021/cb3002478

[15] Nishihara R, Paulmurugan R, Nakajima T, Yamamoto E, Natarajan A, Afjei R, et al. Highly bright and stable NIR-BRET with blue-shifted coelenterazine derivatives for deep-tissue imaging of molecular events in vivo. Theranostics. 2019;9(9):2646–2661. DOI: 10.7150/thno.32219

[16] Anderson N, Anderson N. The human plasma proteome: history, character, and diagnostic prospects. Molecular & Cellular Proteomics. 2002;1(11):845–867. DOI: 10.1074/mcp.r200007-mcp200

[17] Kragh-Hansen U. Molecular and practical aspects of the enzymatic properties of human serum albumin and of albumin-ligand complexes. Biochimica Et Biophysica Acta-General Subjects. 2013;1830(12):5535–5544. DOI: 10.1016/j.bbagen.2013.03.015

[18] Taverna M, Marie A, Mira J, Guidet B. Specific antioxidant properties of human serum albumin. Annals of Intensive Care. 2013;3(1):4. DOI: 10.1186/2110-5820-3-4

[19] Hollfelder F, Kirby A, Tawfik D, Kikuchi K, Hilvert D. Characterization of proton-transfer catalysis by serum albumins. Journal of the American Chemical Society. 2000;122(6):1022–1029. DOI: 10.1021/ja993471y

[20] Hirano T, Gomi Y, Takahashi T, Kitahara K, Qi C, Mizoguchi I, et al. Chemiluminescence of coelenterazine analogs - structures of emitting species. Tetrahedron Letters. 1992;33(39):5771–5774. DOI: 10.1016/0040-4039(92)89028-B

[21] Ikeda Y, Tanaka M, Nishihara R, Hiruta Y, Citterio D, Suzuki K, et al. Quantitative evaluation of luminescence intensity from enzymatic luminescence reaction of coelenterazine and analogues. Journal of Photochemistry and Photobiology A Chemistry. 2020:112459. DOI: 10.1016/j.jphotochem.2020.112459

[22] Ando Y, Niwa K, Yamada N, Enomot T, Irie T, Kubota H, et al. Firefly bioluminescence quantum yield and colour change by pH-sensitive green emission. Nature Photonics. 2008;2(1):44–47. DOI: 10.1038/nphoton.2007.251

[23] Coutant E, Gagnot G, Hervin V, Baatallah R, Goyard S, Jacob Y, et al. Bioluminescence profiling of NanoKAZ/NanoLuc luciferase using a chemical library of coelenterazine analogues. Chemistry-A European Journal. 2020;26(4):948–958. DOI: 10.1002/chem.201904844

[24] Yeh H, Karmach O, Ji A, Carter D, Martins-Green M, Ai H. Red-shifted luciferase-luciferin pairs for enhanced bioluminescence imaging. Nature Methods. 2017;14(10):971–974. DOI: 10.1038/nmeth.4400

[25] Matthews JC, Hori K, Cormier MJ. Substrate and substrate analog binding properties of Renilla luciferase. Biochemistry. 1977;16(24):5217–5220. DOI: 10.1021/bi00643a009

[26] Sugio S, Kashima A, Mochizuki S, Noda M, Kobayashi K. Crystal structure of human serum albumin at 2.5 A resolution. Protein Engineering. 1999;12(6):439–446. DOI: 10.1093/protein/12.6.439

[27] Ghuman J, Zunszain P, Petitpas I, Bhattacharya A, Otagiri M, Curry S. Structural basis of the drug-binding specificity of human serum albumin. Journal of Molecular Biology. 2005;353(1):38–52. DOI: 10.1016/j.jmb.2005.07.075

[28] Xu Y, Su M, Li H, Liu Q, Xu C, Yang Y, et al. A fluorescent sensor for discrimination of HSA from BSA

through selectivity evolution. Analytica Chimica Acta. 2018;1043:123–131. DOI: 10.1016/j.aca.2018.09.010

[29] Murch S, Winyard P, Koletzko S, Wehner B, Cheema H, Risdon R, et al. Congenital enterocyte heparan sulphate deficiency with massive albumin loss, secretory diarrhoea, and malnutrition. Lancet. 1996;347(9011):1299–1301. DOI: 10.1016/s0140-6736(96)90941-1

[30] Luo Z, Lv T, Zhu K, Li Y, Wang L, Gooding J, et al. Paper-based ratiometric fluorescence analytical devices towards point-of-care testing of human serum albumin. Angewandte Chemie-International Edition. 2020;59(8):3131–3136. DOI: 10.1002/anie.201915046

[31] Gao T, Yang S, Cao X, Dong J, Zhao N, Ge P, et al. Smart self-assembled organic nanoprobe for protein-specific detection: design, synthesis, application, and mechanism studies. Analytical Chemistry. 2017;89 (18):10085–10093. DOI: 10.1021/acs.analchem.7b02923

Chapter 3

Near-Infrared Luciferin Analogs for *In Vivo* Optical Imaging

Ryohei Saito-Moriya, Rika Obata and Shojiro A. Maki

Abstract

The firefly bioluminescence reaction has been exploited for *in vivo* optical imaging in life sciences. To develop highly sensitive bioluminescence imaging technology, many researchers have synthesized luciferin analogs and luciferase mutants. This chapter first discusses synthetic luciferin analogs and their structure–activity relationships at the luminescence wavelength of the firefly bioluminescence reaction. We then discuss the development of luciferin analogs that produce near-infrared (NIR) light. Since NIR light is highly permeable for biological tissues, NIR luciferin analogs might sensitively detect signals from deep biological tissues such as the brain and lungs. Finally, we introduce two NIR luciferin analogs (TokeOni and seMpai) and a newly developed bioluminescence imaging system (AkaBLI). TokeOni can detect single-cell signals in mouse tissue and luminescence signals from marmoset brain, whereas seMpai can detect breast cancer micro-metastasis. Both reagents are valid for *in vivo* bioluminescence imaging with high sensitivity.

Keywords: Firefly bioluminescence, Bioluminescence imaging, Structure–activity relationships, Multicolor, Near-infrared light

1. Introduction

In Japan, watching the light of fireflies has been a summer tradition for over one thousand years. Modern fireflies are known to glow yellow-green, but in ancient times they emitted a dark green luminescence, as confirmed by recent molecular biology techniques [1]. The detailed mechanism of firefly bioluminescence is described in previous chapters. This chapter focuses on synthetic substrates of firefly luciferase, which are employed in firefly bioluminescence imaging (BLI).

In recent biological research, BLI technology has observed biological events *in vivo* [2–8]. For example, in cancer research, BLI has been applied to real-time monitoring of gene expression, cell numbers, and other biological events in transgenic mouse models [9–16]. Our group has developed firefly substrate analogs for use in these research fields.

The firefly bioluminescence reaction proceeds via the oxidation of D-luciferin (**1**, LH2, **Figure 1**) catalyzed by firefly luciferase (Fluc) in the presence of adenosine triphosphate (ATP), Mg^{2+} and O_2 by a two-step reaction. In the first step, LH2 is adenylated with ATP, and is then oxidized by O_2, forming excited-state oxyluciferin that relaxes to the ground state with yellow-green light emission (λ_{max} = 560 nm) [17–19]. However, yellow-green light is not able to easily penetrate biological tissues [20], and is useful only for imaging shallow tissues such as subcutaneous tissues. To detect signals from deep tissues such as brain and lung [21], near-infrared (NIR)

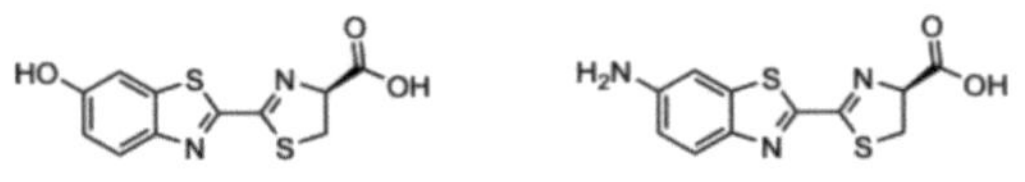

Figure 1.
Structures of D-luciferin (1, LH2) and aminoluciferin (2).

light should be used, as it is highly permeable to biological tissues [20] and is suitable for *in vivo* deep tissue imaging [21]. Recently, many synthetic luciferin analogs have been reported. Our group has synthesized various luciferin analogs and compared them with **1**. By studying the structure–activity relationships of these analogs and *Photinus pyralis* (*Ppy*) luciferase, we have developed luciferin analogs that produce wide-spectrum light (from blue to red), along with NIR luciferin analogs (AkaLumine, TokeOni, seMpai) for BLI. Our different analogs are described in this chapter.

2. Luciferin analogs of firefly luciferase

Many researchers have synthesized luciferin analogs, and different substrates reacted with luciferases exhibit different luminescence activities [22–24]. Most luciferin analogs are formed by modifying the benzothiazole moiety of **1**. Analogs of **1** were first synthesized by White *et al.* in 1966. They showed that aminoluciferin (2, **Figure 1**), in which the hydroxyl group of benzothiazole is replaced with an amino group, can function as a substrate of Fluc and emit red bioluminescence [25].

2.1 Development of luciferin analogs based on LH2

As mentioned above, many luciferin analogs are prepared by modifying the benzothiazole moiety of **1** [22–24]. For instance, *N*-cycloaminoluciferins (**2a–f**, **Figure 2A**) are prepared by cyclizing the NH_2 of **2**. These analogs were reported by two independent groups, who synthesized them by different routes [26, 27] (**Figure 2B–C**). When reacted with Fluc, **2a–f** show longer wavelengths than **1**, probably reflecting the electron donation effect of cycloamine substitutes. Comparing the bioluminescence activities and emission wavelengths of analogs **2e** and **2f** on Fluc and Fluc mutant luciferase R218K, it was found that **2e**/Fluc and **2f**/R218K produced light at 604 and 614 nm, respectively, whereas **2e**/R218K and **2f**/Fluc produced no light [26]. The interaction between the active site of luciferase and the substrate is very critical, indicating that the structures of both reactants play essential controlling roles in luminescence activity.

Miller *et al.* synthesized CycLuc1 (**7a**, **Figure 3**) by fusing *N*-cycloalikylation of **2** with benzothiazole [28]. Analog **7a** exhibited a longer luminescence wavelength on Fluc (599 nm) than **1** on Fluc, and was emitted more intensely than **1** in a *Photuris pennsylvanica* firefly luciferase mutant (Ultra-Glo). The BLI of **7a** detects the signals from deep organs such as brains and lungs [21, 29]. Li *et al.* synthesized CybLuc (**7b**, **Figure 3**) by substituting the hydroxy group of **2** with a cycloamino group. Analog **7b** produced light at 603 nm and its BLI detected the signals from mouse brain [30].

Iwano *et al.* developed luciferin analogs **8a–g** (**Figure 4A**) by substituting the benzothiazole moiety of **1** with a simple benzene ring and extended π-conjugations [31]. Olefins were extended by the Wittig reaction from **10c–d** and **12e–f** as starting

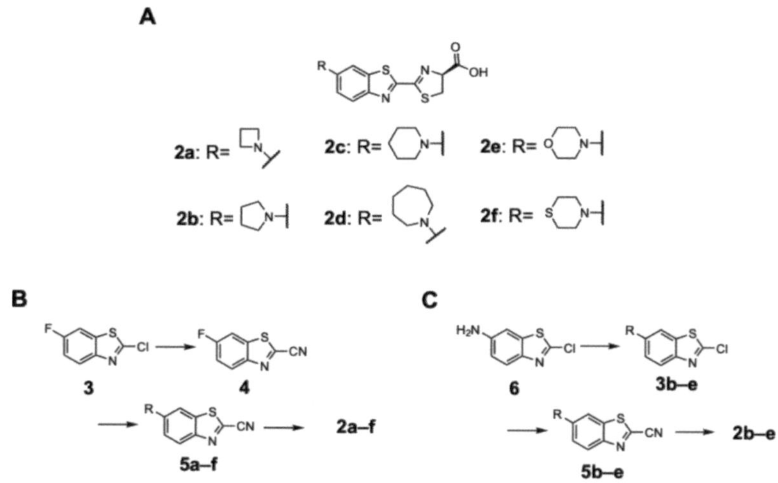

Figure 2.
*Structures and synthetic routes of luciferin analogs **2a–f**. (**A**) N-cycloaminoluciferin analogs **2a–f**, (**B**) the synthetic route reported by Miller et al. [28], and (**C**) the synthetic route reported by Hirano et al. [29].*

CycLuc1 (7a)   CybLuc (7b)

Figure 3.
*Structures of CycLuc1 (**7a**) and CybLuc (**7b**).*

materials. In this synthesis, hydrolysis was stepwise followed by condensation with D-Cys(STrt)-OMe, thiazoline cyclization, and methyl ester deprotection (**Figure 4B**). The obtained analogs **8a–f** produced luminescent colors over a wide range (blue to red) [31]. Among these, AkaLumine (**8e**), which produces light at 675 nm, is a leading compound for NIR luciferin analogs, as described in Section 2.2. Later, analog **8e** was used as a reagent for BLI. In the same paper, 3-hydroxyl analog **8g** (**Figure 4A**) was also synthesized, but this analog produced no light [31]. Therefore, the position of the OH substituent is critical in the firefly bioluminescence reaction.

In contrast, the thiazoline site is rarely modified. Conley *et al*. synthesized a seleno-aminoluciferin analog **13a** (**Figure 5A**) in which the S of the thiazoline ring of **2** was replaced with Se [32], and Ioka *et al*. synthesized O- or C- substituted analogs **13b–c** (**Figure 5A**) [33]. Analog **13a**, which produced light at 600 nm, was synthesized by the cyclization reaction of selenocysteine (**Figure 5B**) [32]. Analog **13b** was obtained by synthesizing an amide **16b** synthesizing an amide from D-serine, cyclizing it with diethylaminosulfur trifluoride (DAST), and hydrolyzing it with Amano lipase (**Figure 5C**). Analog **13c** was prepared by coupling with bromothiazole **19** and pyrrolidione carboxylate **21** to form glutamate-linked benzothiazole **16c**, and cyclizing **16c** with trifluoroacetic acid (TFA) (**Figure 5D**). Interestingly, **13c** produces light at 547 nm, whereas **13b** is non-bioluminescence [33] but shows chemiluminescent ability. This result indicates that the thiazoline of **1** is an essential moiety for recognizing the activity site in luciferase.

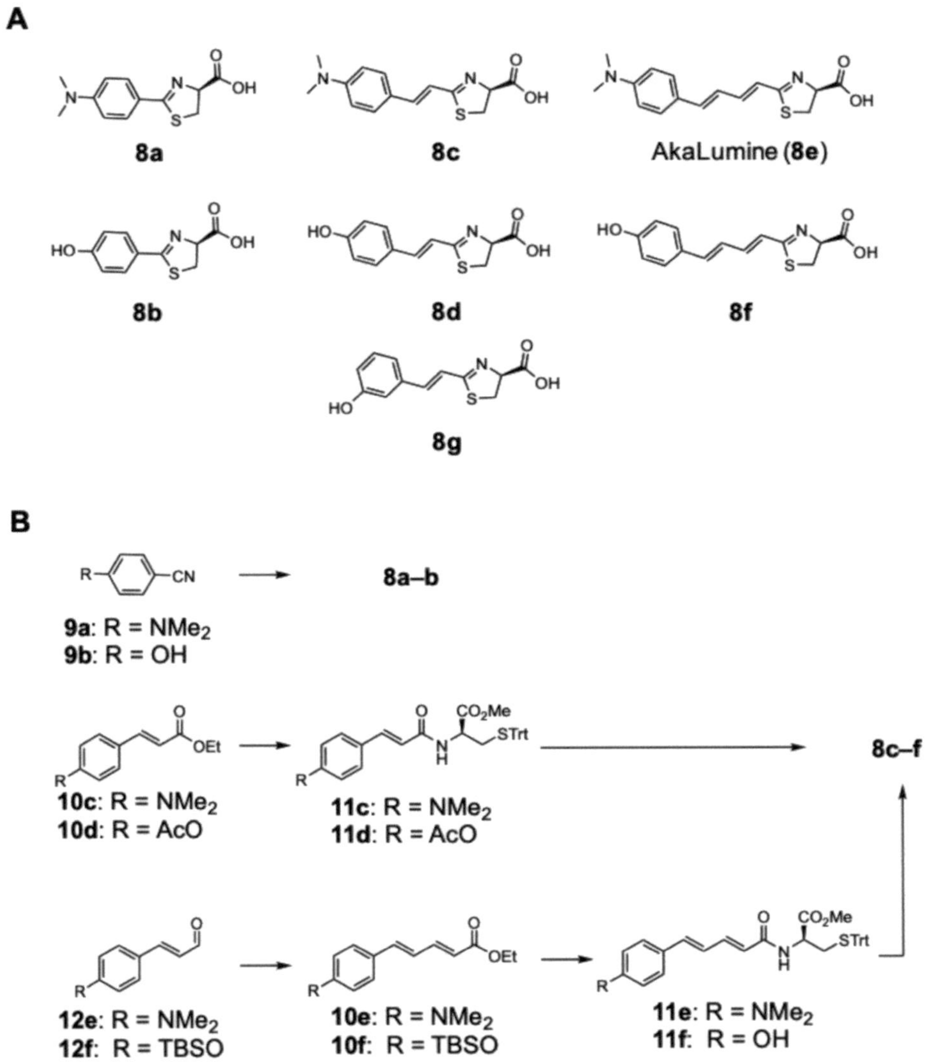

Figure 4.
*Structures of **8a–g** (**A**) and their synthetic routes **8a–f** (**B**).*

2.2 Structure–activity relationships for developing NIR luciferin analogs

Based on these structure–activity relationships, additional luciferin analogs have been designed and synthesized for NIR light production. For example, Anderson *et al*. synthesized iLH2 (**22**, **Figure 6**) by inserting an olefin into the structure of **1**. Analog **22** produced NIR light at 706 nm [34]. However, the luciferase used at that time was a mutant (S284T), and the luminescence wavelength on Fluc was 670 nm. The same authors developed an *in vivo* dual-imaging technique that combines **1** and **22** with two different luciferases. This system can potentially observe new biological events by tracking two processes simultaneously [35]. Hall *et al*. synthesized NH_2–NpLH2 (**23**, **Figure 6**) by extended conjugation of **2**. Analog **23** produced no light with Fluc, but its luminescence wavelength was extended to 743 nm by reaction with CBR2, a mutant luciferase of click beetles (*Pyrophorus plagiophthalamus*) [36]. All of these studies achieved long-wavelength emissions from mutant luciferases, but their luminescence activity is much lower than that of combinations of **1** and wild type *Ppy* luciferase.

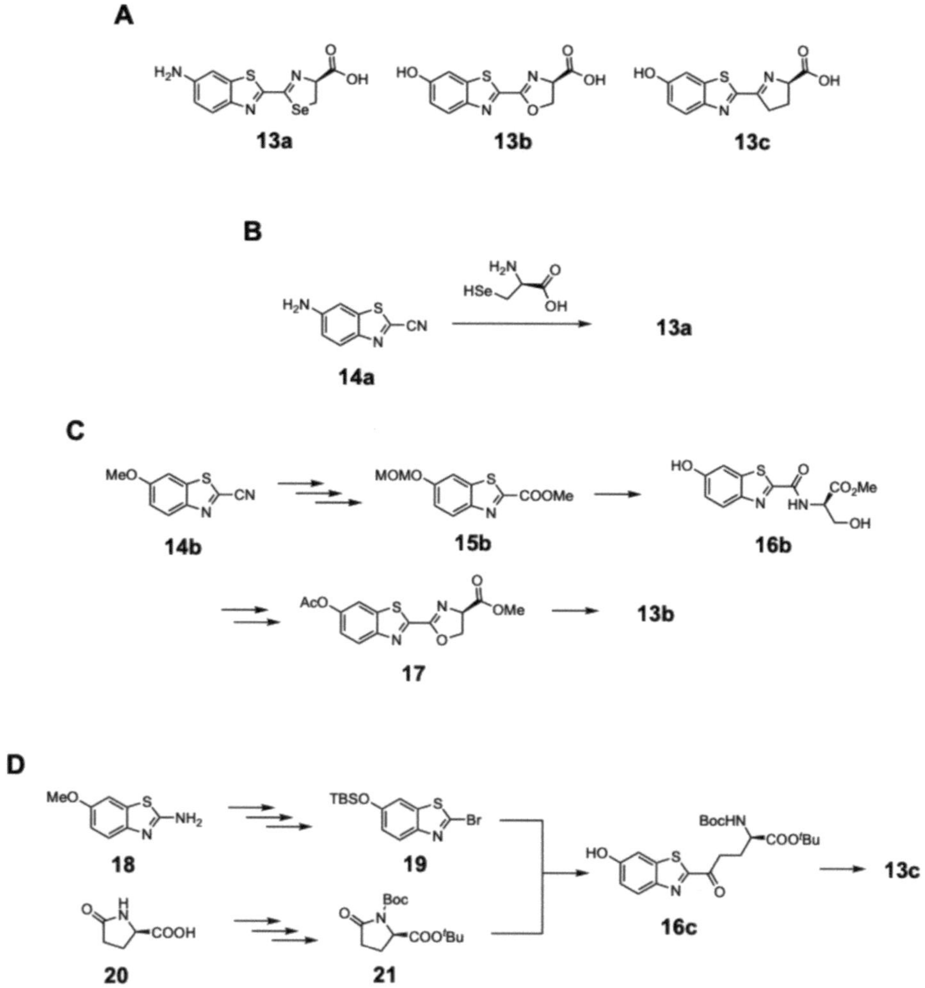

Figure 5.
Structures of ***13a–c*** *(A) and their synthetic routes* ***13a*** *(B),* ***13b*** *(C) and* ***13c*** *(D).*

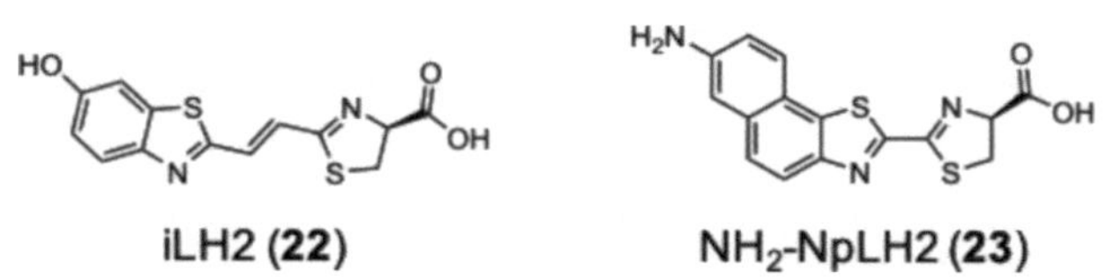

Figure 6.
Structures of NIR luciferin analogs iLH2 (22) and NH_2-NpLH2 (23).

Meanwhile, Maki's group has developed a number of analogs based on the structure of **8e**, which are expected to produce NIR light. Miura *et al*. formed a mother skeleton by a coupling reaction, and thus synthesized biphenyl analogs **24a–c** (**Figure 7** and **8A**) [37]. Analog **24a** produced light at 675 nm, but the luminescence intensity was weak. Although its conjugation was more extended than in **8e**, the luminescence wavelength of **24a** did not change as that of **8e** (675 nm). This result suggests that the biphenyl moiety rotates and reduces the fluorescent intensity.

Kiyama *et al*. synthesized cyclic amino analogs of **8e (25a–d, Figure 7)** [38] from 4-fluorobenzaldehyde **32** as the starting material. They replaced the F group with various secondary amines, and conducted the Horner–Wadsworth–Emmons

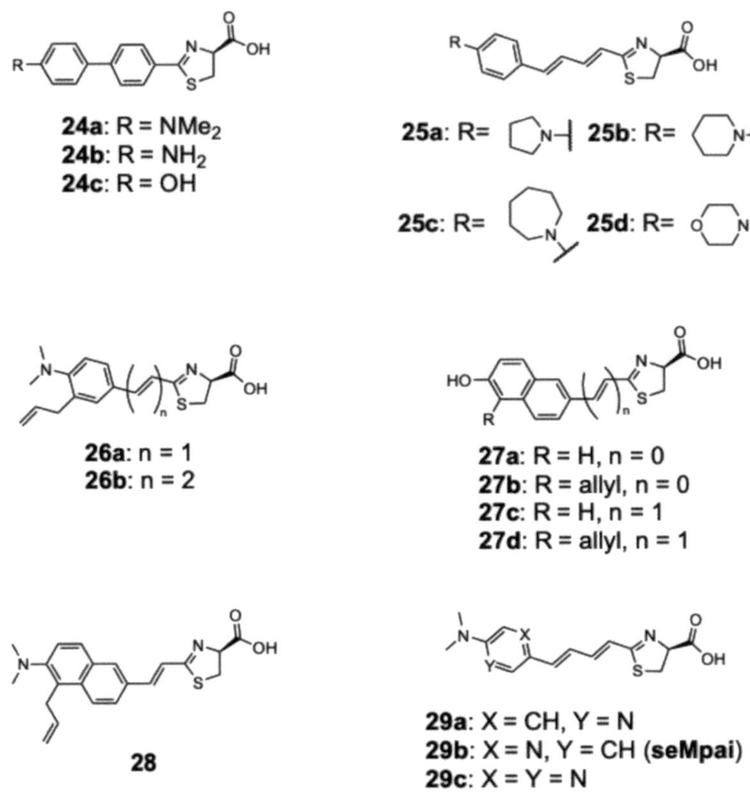

Figure 7.
Structures of NIR luciferin analogs **24–29**.

reaction, condensation and cyclization to obtain the final compounds **25a–d** (**Figure 8B**). Despite containing an electron-donating amino group, **25a–d** produced luminescence at almost the same wavelengths (656–667 nm) as **8e** (668 nm). However, the luminescence intensity of **25a** was approximately four times stronger than that of **8e**. The fluorescence quantum yields of **8e** and these cyclic amino analogs **25a–d** were almost identical, suggesting that the luminescence intensity largely depends on the reactivity with luciferase.

The luminescent wavelength can be lengthened not only by extending the π-conjugations and introducing an electron donate substituent, but also by introducing an allyl group. Kitada *et al.* synthesized allyl analogs **26a–b** (**Figure 7**) by introducing allyl groups into **8c**, **8e** and naphthol analogs **27a–d** (**Figure 7**). The analogs were introduced by two routes: Pd-catalyzed Stille coupling (**Figure 8C**) and Claisen rearrangement (**Figure 8D**) [39]. Although these analogs delivered very low luminescence intensities, their wavelength shift was long (approximately 15–35 nm). As the allyl group itself does not affect the π-conjugations of the substrate structure, it was considered that induce fitting was occurred at the luciferase active site and stabilized the substrate metabolite to lower energy state conformation. To develop a long-wavelength, Kitada *et al.* synthesized NIR analog (**28** in **Figure 7**) by introduced both an electron-donating NMe_2 and an allyl group. When reacted with Fluc, **28** produced NIR light at a sufficiently long-wavelength (705 nm), but the luminescence intensity was only 1.3% of that of **8e**. Although the allyl group extends the luminescent wavelength, it greatly reduces the luminescence intensity, which is a major disadvantage.

The aromatic ring site has also been targeted in the development of potential NIR emitters. Saito *et al.* synthesized three analogs **29a–c** (**Figure 7**) in which the

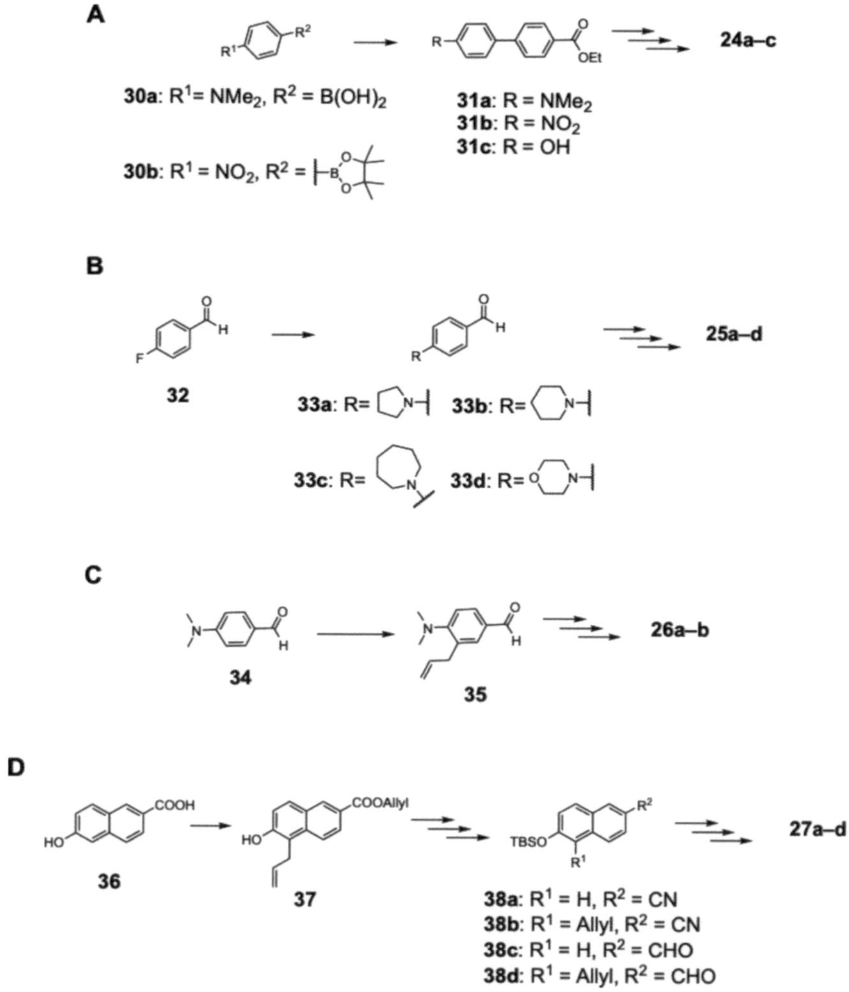

Figure 8.
Synthetic routes of NIR luciferin analogs **24a–c** *(A),* **25a–d** *(B),* **26a–b** *(C) and* **27a–d** *(D).*

aromatic ring of **8e** was replaced with an *N*-heteroaromatic ring [40]. Interestingly, the luminescence wavelengths of three analogs depended on the positions and numbers of their N atoms; **29a** produced red light at 645 nm, seMpai (**29b**) produced NIR light at 675 nm, and **29c** produced orange light at 625 nm. This result highlights the importance of interactions between the luciferase active site and the N atoms of the heterocycle. Although the luminescence wavelength of all three analogs were shorter than 700 nm, the wavelength was changed with a single atom, suggesting that interaction with the luciferase active site is an important part of molecular design.

3. Solubility enhancement of TokeOni and seMpai (29b) for sensitive *in vivo* imaging

As mentioned in the previous section, our research group has developed multicolor luciferin analogs for *in vivo* imaging. The luminescence activities of AkaLumine (**8e**) and seMpai (**29b**) are potentially suitable for BLI. Therefore, the usefulness of these analogs as reagents for *in vivo* BLI was evaluated in further animal experiments.

3.1 Development of AkaBLI (TokeOni and AkaLuc)

Prior to administering the reagent into the animal models, we increased the aqueous solubility of **8e** (which is inherently low) and developed an HCl salt of **8e**, AkaLumine-HCl (Tokeoni) [21]. In ultrapure water, the solubility of TokeOni was 40 mM, approximately 20 times higher than that of **8e** (2 mM). Therefore, TokeOni enable to administered to experimental animals with a smaller solution volume and higher concentration than **8e**. TokeOni/Fluc BLI was performed with significantly higher sensitivity than LH2/Fluc and CycLuc1/Fluc in the lungs [21] and brain [41] of mice. However, as the *in vitro* luminescence intensity of TokeOni/Fluc was lower than that of LH2/Fluc, it must be improved before imaging large animals such as marmosets.

Accordingly, Iwano *et al*. developed a mutant luciferase Akaluc specialized for TokeOni. They developed the artificial bioluminescence system AkaBLI, which combines TokeOni with Akaluc [42] (a mutation of 28 amino acid residues on *Ppy* luciferase). The AkaBLI luminescence intensity was approximately 10-fold higher in cells (*in vitro*), 52-fold higher in mouse lungs (*in vivo*), and 1400-fold higher in mouse brain tissue (*in vivo*) than LH2/Fluc luminescence intensity. In addition, AkaBLI detected single-cell signals from mouse lung and to quantified 1–10 cells. For large animal imaging, the authors inserted the Akaluc gene into an adeno-associated virus (AAV) vector, and introduced the recombinant AVV into the striatal neurons of marmosets. AkaBLI achieved video-rate real-time imaging of marmoset brains.

3.2 BLI with seMpai

seMpai (**29b**) was developed to improve two weak points in TokeOni: neutral pH and disturbance by hepatic background signals. As TokeOni is acidic [43], it may cause acidosis when injected; moreover, TokeOni detects the hepatic background signals, which are not detected by LH2 [43, 44]. Due to the effect of N atom, seMpai was sufficiently soluble for administration to experimental animals and 69 mM was dissolved in phosphate buffered saline (pH 7.4) [40]. In Fluc-expressing lung cancer model mice, the sensitivities of seMpai and TokeOni were not significantly different [40], but seMpai detected no hepatic background signals and seMpai BLI detected breast cancer micro-metastasis [43]. When repeated with TokeOni and LH2, this experiment was unsuccessful. Although seMpai/Fluc was less sensitive than AkaBLI for single-cell imaging, its imaging sensitivity could be improved mutant luciferases such as Akaluc.

Fukuchi *et al*. monitored the expression of brain-derived neurotrophic factor (BDNF) in Bdnf-luc transgenic mice with LH2, TokeOni, and seMpai [45]. TokeOni achieved the most sensitive BLI, and seMpai and LH2 were comparable. The result probably reflects the different abilities of the compounds to penetrate the blood–brain-barrier. This result also indicates the necessary of evaluating the imaging reagent in terms of both its luminescence activity and pharmacokinetics.

Additionally, biocompatibility such as cytotoxicity of TokeOni and seMpai has not been reported yet. No acute toxicity or adverse side effects were observed in mice when these compounds were administered at a concentration of 33 mM [40], however, preliminary experiments are recommended when using new analogs, not limited to TokeOni and seMpai.

4. Conclusion

By investigating the structure–activity relationship of luciferin analogs, researchers have developed various methods for tuning the luminescence wavelengths of these analogs. However, the intensity of the luminescence is poorly controlled. If the luminescence intensity and structure–activity relationships could be to associated by a predictable law, we could synthesize new luciferin analogs with high luminescence intensity, and further develop an imaging technology with greater usefulness than conventional technologies.

For a practical imaging technology, both the luminescent substrate/enzyme activity and the pharmacokinetics are very important. Improving the various properties of the substrates and enzymes will enhance the sensitivity of bioluminescence imaging.

Acknowledgements

We thank the members of Maki's and Aoyama's laboratory, and Kurogane Kasei Co., Ltd. for meaningful discussions and support.

Conflict of interest

The authors declare no conflict of interest.

Author details

Ryohei Saito-Moriya[1,2,3], Rika Obata[1] and Shojiro A. Maki[1,2]*

1 Department of Engineering and Science, Graduate School of Informatics and Engineering, The University of Electro-Communications, Tokyo, Japan

2 Center for Neuroscience and Biomedical Engineering, The University of Electro-Communications, Tokyo, Japan

3 School of Pharmacy, Tokyo University of Pharmacy and Life Sciences, Tokyo, Japan

*Address all correspondence to: s-maki@uec.ac.jp

References

[1] Oba Y, Konishi K, Yano D, Shibata H, Kato D, Shirai T. Resurrecting the ancient glow of the fireflies. Sci Adv. 2020;6(49):eabc5705. DOI: 10.1126/sciadv.abc5705

[2] Contag PR, Olomu IN, Stevenson DK, Contag CH. Bioluminescent indicators in living mammals. Nat Med. 1998;4(2):245-7. DOI: 10.1038/nm0298-245h

[3] Kim J-B, Urban K, Cochran E, Lee S, Ang A, Rice B, et al. Non-invasive detection of a small number of bioluminescent cancer cells in vivo. PLoS One. 2010;5(2):e9364. DOI: 10.1371/journal.pone.0009364

[4] Badr CE, Tannous BA. Bioluminescence imaging: progress and applications. Trends Biotechnol. 2011;29(12):624-33. DOI: 10.1016/j.tibtech.2011.06.010

[5] Kaskova ZM, Tsarkova AS, Yampolsky I V. 1001 lights: luciferins, luciferases, their mechanisms of action and applications in chemical analysis, biology and medicine. Chem Soc Rev. 2016;45(21):6048-77. DOI: 10.1039/C6CS00296J

[6] Love AC, Prescher JA. Seeing (and using) the light: Recent developments in bioluminescence technology. Cell Chem Biol. 2020;27(8):904-20. DOI: 10.1016/j.chembiol.2020.07.022

[7] Chen L, Chen Z, Zheng S, Fan L, Zhu L, Yu J, et al. Study on mechanism of elemene reversing tumor multidrug resistance based on luminescence pharmacokinetics in tumor cells in vitro and in vivo. RSC Adv. 2020;10(57):34928-37. DOI: 10.1039/D0RA00184H

[8] Yamada K, Noguchi K, Kimitsuki K, Kaimori R, Saito N, Komeno T, et al. Reevaluation of the efficacy of favipiravir against rabies virus using in vivo imaging analysis. Antiviral Res. 2019;172:104641. DOI: 10.1016/j.antiviral.2019.104641

[9] Bennett J, Duan D, Engelhardt JF, Maguire AM. Real-time, noninvasive in vivo assessment of adeno-associated virus-mediated retinal transduction. Invest Ophthalmol Vis Sci. 1997;38(13):2857-63. PMID: 9418740

[10] Yanagihara K, Takigahira M, Takeshita F, Komatsu T, Nishio K, Hasegawa F, et al. A photon counting technique for quantitatively evaluating progression of peritoneal tumor dissemination. Cancer Res. 2006;66(15):7532-9. DOI: 10.1158/0008-5472.CAN-05-3259

[11] Nakayama J, Ito E, Fujimoto J, Watanabe S, Semba K. Comparative analysis of gene regulatory networks of highly metastatic breast cancer cells established by orthotopic transplantation and intra-circulation injection. Int J Oncol. 2017;50(2):497-504. DOI: 0.3892/ijo.2016.3809

[12] Kuchimaru T, Kataoka N, Nakagawa K, Isozaki T, Miyabara H, Minegishi M, et al. A reliable murine model of bone metastasis by injecting cancer cells through caudal arteries. Nat Commun. 2018;9(1):2981. DOI: 10.1038/s41467-018-05366-3

[13] Fleiss A, Sarkisyan KS. A brief review of bioluminescent systems (2019). Curr Genet. 2019;65(4):877-82. DOI: 10.1007/s00294-019-00951-5

[14] Manni I, de Latouliere L, Gurtner A, Piaggio G. Transgenic animal models to visualize cancer-related cellular processes by bioluminescence imaging. Front Pharmacol. 2019;10(March):1-12. DOI: 10.3389/fphar.2019.00235

[15] Kuroiwa Y, Nakayama J, Adachi C, Inoue T, Watanabe S, Semba K. Proliferative classification of intracranially injected HER2-positive breast cancer cell lines. Cancers (Basel). 2020;12(7):1811. DOI: 10.3390/cancers12071811

[16] Han Y, Nakayama J, Hayashi Y, Jeong S, Futakuchi M, Ito E, et al. Establishment and characterization of highly osteolytic luminal breast cancer cell lines by intracaudal arterial injection. Genes to Cells. 2020;25(2):111-23. DOI: 10.1111/gtc.12743

[17] Branchini BR, Murtiashaw MH, Magyar RA, Portier NC, Ruggiero MC, Stroh JG. Yellow-green and red firefly bioluminescence from 5,5-dimethyloxyluciferin. J Am Chem Soc. 2002;124(10):2112-3. DOI: 10.1021/ja017400m

[18] Branchini BR, Southworth TL, Murtiashaw MH, Magyar RA, Gonzalez SA, Ruggiero MC, et al. An alternative mechanism of bioluminescence color determination in firefly luciferase. Biochemistry. 2004(23):7255-62. DOI: 10.1021/bi036175d

[19] Hirano T, Hasumi Y, Ohtsuka K, Maki S, Niwa H, Yamaji M, et al. Spectroscopic studies of the light-color modulation mechanism of firefly (beetle) bioluminescence. J Am Chem Soc. 2009;131(6):2385-96. DOI: 10.1021/ja808836b

[20] Weissleder R. A clearer vision for in vivo imaging. Nat Biotechnol. 2001 (4):316-7. DOI: 10.1038/86684

[21] Kuchimaru T, Iwano S, Kiyama M, Mitsumata S, Kadonosono T, Niwa H, et al. A luciferin analogue generating near-infrared bioluminescence achieves highly sensitive deep-tissue imaging. Nat Commun. 2016;7(1):11856. DOI: 10.1038/ncomms11856

[22] Miller SC, Mofford DM, Adams ST. Lessons learned from luminous luciferins and latent luciferases. ACS Chem Biol. 2018;13(7):1734-40. DOI: 10.1021/acschembio.7b00964

[23] Podsiadły R, Grzelakowska A, Modrzejewska J, Siarkiewicz P, Słowiński D, Szala M, et al. Recent progress in the synthesis of firefly luciferin derivatives. Dye Pigment. 2019;170(February):107627. DOI: 10.1016/j.dyepig.2019.107627

[24] Li S, Ruan Z, Zhang H, Xu H. Recent achievements of bioluminescence imaging based on firefly luciferin-luciferase system. Eur J Med Chem. 2021;211:113111. DOI: 10.1016/j.ejmech.2020.113111

[25] White EH, Wörther H, Seliger HH, McElroy WD. Amino analogs of firefly luciferin and biological activity thereof. J Am Chem Soc. 1966;88(9):2015-9. DOI: 10.1021/ja00961a030

[26] Sharma DK, Adams ST, Liebmann KL, Miller SC. Rapid access to a broad range of 6′-substituted firefly luciferin analogues reveals surprising emitters and inhibitors. Org Lett. 2017;19(21):5836-9. DOI: 10.1021/acs.orglett.7b02806

[27] Kakiuchi M, Ito S, Kiyama M, Goto F, Matsuhashi T, Yamaji M, et al. Electronic and steric effects of cyclic amino substituents of luciferin analogues on a firefly luciferin–luciferase reaction. Chem Lett. 2017;46(8):1090-2. DOI: 10.1246/cl.170361

[28] Reddy GR, Thompson WC, Miller SC. Robust light emission from cyclic alkylaminoluciferin substrates for firefly luciferase. J Am Chem Soc. 2010;132(39):13586-7. DOI: 10.1021/ja104525m

[29] Evans MS, Chaurette JP, Adams ST, Reddy GR, Paley MA, Aronin N, et al.

A synthetic luciferin improves bioluminescence imaging in live mice. Nat Methods. 2014;11(4):393-5. DOI: 10.1038/nmeth.2839

[30] Wu W, Su J, Tang C, Bai H, Ma Z, Zhang T, et al. cybLuc: An effective aminoluciferin derivative for deep bioluminescence imaging. Anal Chem. 2017;89(9):4808-16. DOI: 10.1021/acs.analchem.6b03510

[31] Iwano S, Obata R, Miura C, Kiyama M, Hama K, Nakamura M, et al. Development of simple firefly luciferin analogs emitting blue, green, red, and near-infrared biological window light. Tetrahedron. 2013;69(19):3847-56. DOI: 10.1016/j.tet.2013.03.050

[32] Conley NR, Dragulescu-Andrasi A, Rao J, Moerner WE. A selenium analogue of firefly D-luciferin with red-shifted bioluminescence emission. Angew Chemie Int Ed. 2012;51(14):3350-3. DOI: 10.1002/anie.201105653

[33] Ioka S, Saitoh T, Iwano S, Suzuki K, Maki SA, Miyawaki A, et al. Synthesis of firefly luciferin analogues and evaluation of the luminescent properties. Chem - A Eur J. 2016;22(27):9330-7. DOI: 10.1002/chem.201600278

[34] Jathoul AP, Grounds H, Anderson JC, Pule MA. A dual-color far-red to near-infrared firefly luciferin analogue designed for multiparametric bioluminescence imaging. Angew Chemie Int Ed. 2014;53(48):13059-63. DOI: 10.1002/anie.201405955

[35] Stowe CL, Burley TA, Allan H, Vinci M, Kramer-Marek G, Ciobota DM, et al. Near-infrared dual bioluminescence imaging in mouse models of cancer using infraluciferin. eLife. 2019;8:1-22. DOI: 10.7554/eLife.45801

[36] Hall MP, Woodroofe CC, Wood MG, Que I, van't Root M, Ridwan Y, et al. Click beetle luciferase mutant and near infrared naphthyl-luciferins for improved bioluminescence imaging. Nat Commun. 2018;9(1):132. DOI: 10.1038/s41467-017-02542-9

[37] Miura C, Kiyama M, Iwano S, Ito K, Obata R, Hirano T, et al. Synthesis and luminescence properties of biphenyl-type firefly luciferin analogs with a new, near-infrared light-emitting bioluminophore. Tetrahedron. 2013;69(46):9726-34. DOI: 10.1016/j.tet.2013.09.018

[38] Kiyama M, Iwano S, Otsuka S, Lu SW, Obata R, Miyawaki A, et al. Quantum yield improvement of red-light-emitting firefly luciferin analogues for in vivo bioluminescence imaging. Tetrahedron. 2018;74(6):652-60. DOI: 10.1016/j.tet.2017.11.051

[39] Kitada N, Saitoh T, Ikeda Y, Iwano S, Obata R, Niwa H, et al. Toward bioluminescence in the near-infrared region: Tuning the emission wavelength of firefly luciferin analogues by allyl substitution. Tetrahedron Lett. 2018;59(12):1087-90.DOI: 10.1016/j.tetlet.2018.01.078

[40] Saito R, Kuchimaru T, Higashi S, Lu SW, Kiyama M, Iwano S, et al. Synthesis and luminescence properties of near-infrared N-heterocyclic luciferin analogues for in vivo optical imaging. Bull Chem Soc Jpn. 2019;92(3):608-18. DOI: 10.1246/bcsj.20180350

[41] Fukuchi M, Izumi H, Mori H, Kiyama M, Otsuka S, Maki S, et al. Visualizing changes in brain-derived neurotrophic factor (BDNF) expression using bioluminescence imaging in living mice. Sci Rep. 2017;7(1):4949. DOI: 10.1038/s41598-017-05297-x

[42] Iwano S, Sugiyama M, Hama H, Watakabe A, Hasegawa N, Kuchimaru T, et al. Single-cell bioluminescence imaging of deep tissue in freely moving animals. Science.

2018;359(6378):935-9. DOI: 10.1126/science.aaq1067

[43] Nakayama J, Saito R, Hayashi Y, Kitada N, Tamaki S, Han Y, et al. High sensitivity in vivo imaging of cancer metastasis using a near-infrared luciferin analogue seMpai. Int J Mol Sci. 2020;21(21):7896. DOI: 10.3390/ijms21217896

[44] Su Y, Walker JR, Park Y, Smith TP, Liu LX, Hall MP, et al. Novel NanoLuc substrates enable bright two-population bioluminescence imaging in animals. Nat Methods. 2020;17(8):852-60. DOI: 10.1038/s41592-020-0889-6

[45] Fukuchi M, Saito R, Maki S, Hagiwara N, Nakajima Y, Mitazaki S, et al. Visualization of activity-regulated BDNF expression in the living mouse brain using non-invasive near-infrared bioluminescence imaging. Mol Brain. 2020;13(1):122. DOI: 10.1186/s13041-020-00665-7

Chapter 4

Imaging Promoter Assay of Adenylyl Cyclase A Gene in *Dictyostelium discoideum* during Fruiting Body Formation by Dual-Color Bioluminescence Microscopy

Taro Hayashi, Katsunori Ogoh and Hirobumi Suzuki

Abstract

Cyclic adenosine monophosphate (cAMP), which is derived from adenosine triphosphate through adenylyl cyclase A (acaA), acts as an intracellular secondary messenger and an extracellular chemotactic substance in important biological processes. In the social amoebae *Dictyostelium discoideum*, cAMP mediates cell aggregation, development, and differentiation to spore and stalk cells during fruiting body formation. The *acaA* gene is transcribed under the control of three different alternative promoters. This study aimed to develop a promoter assay for *acaA* in *D. discoideum* using bioluminescence microscopy. Here, we inserted green- and red-emitting luciferase genes into downstream of promoter regions 1 and 3, respectively. Promoter activities were visualized by bioluminescence microscopy. We confirmed the differential expression of *acaA* under the control of promoters 1 and 3 at the different stages of *D. discoideum* development. We also demonstrated the application of dual-color bioluminescence imaging in the development of an imaging promoter assay.

Keywords: *Dictyostelium discoideum*, adenylyl cyclase A promoter, dual-color luciferase, bioluminescence microscopy, imaging promoter assay, fruiting body formation

1. Introduction

Gene expression and regulation are essential processes in cellular proliferation and differentiation and are involved in morphogenesis and embryogenesis. The social amoebae *Dictyostelium discoideum* is known to have a simple life cycle, short generation time, and small genome size. Thus, it is a model organism that is often used in research on morphogenesis, especially the formation of fruiting bodies from amoeba cell aggregation. This cell aggregation is mediated by extracellular cyclic adenosine monophosphate (cAMP) [1]. Then, cAMP is secreted from the tip of the cell mound for prestalk and prespore cell migration [2]. High concentrations of extracellular cAMP are required for prestalk and prespore cell differentiation [3, 4] and spore

IntechOpen

formation [5]. Thus, cAMP plays an important role in *Dictyostelium discoideum* development. Its synthesis is catalyzed by adenylyl cyclases A, B, and G, which are encoded by the genes *acaA*, *acrA*, and *acgA*, respectively [6, 7].

Adenylyl cyclase A is considered a development-specific enzyme [8, 9]. Galardi-Castilla et al. [10] characterized the promoter region of the *acaA* gene in *Dictyostelium discoideum* Ax4 cells by histochemistry using a *lacZ*/X-Gal staining system, β-galactosidase reporter system, quantitative RT-PCR, and *in situ* hybridization. The *acaA* gene is transcribed under the control of three different alternative promoters: promoter 1 (distal region), promoter 2 (intermediate region), and promoter 3 (proximal region). Promoter 1 is active during the cell aggregation stage, and promoters 2 and 3 are active in the mound, slug, and fruiting body stages [10].

Promoter assays using histochemical techniques are quite cumbersome, as the samples have to be fixed, stained, and observed over time sequentially for each promoter. For this purpose, many samples must be prepared. On the other hand, a promoter assay using bioluminescence microscopy can be used to obtain time-lapse image data from a single experiment using one sample. This method is often used in the study of clock genes [11–13] and developmental biology [14–16]. In this chapter, we applied bioluminescence microscopy and used two luciferases in the development of an *acaA* promoter assay that can monitor *acaA* promoters 1 and 3 simultaneously during *Dictyostelium discoideum* development, and compared the result with those of histochemistry and β-galactosidase reporter system [10]. We also demonstrated the advantages of this promoter assay and discussed the perspectives that need further consideration.

2. Materials and methods

2.1 Dictyostelium discoideum

Under the National BioResource Project (NBRP), the National Institute of Advanced Industrial Science and Technology (AIST) in Japan provided the *Dictyostelium discoideum* strain Ax2 (NBRP ID: S00001). The Ax2 cells were cultured in SM/5 medium on a 1.4% agar plate at 21°C with *Klebsiella aerogenes* bacterial cells (provided by Prof. H. Kuwayama, Tsukuba University, Japan) as feed.

2.2 Firefly luciferase gene

Green- and red-emitting luciferases were used for the dual-color bioluminescence promoter assay. The green-emitting luciferase gene *Luci sp1* was cloned from *Luciola* sp. collected in the Belum forest, State Park, Malaysia. *Luci sp1* was modified and optimized for mammalian cell expression. Variant 1 of *Luci sp1* [17] (DNA Data Bank of Japan [DDBJ] accession no. LC632706) was used for the green vector construction. The red-emitting luciferase gene *Psa* (wild-type) was cloned from *Pristolycus sagulatus* collected in Tokyo, Japan. *Psa* (Wildtype) was modified and optimized for mammalian cell expression as *Psa* [18] (DDBJ accession no. LC495933). This luciferase gene was used for the red vector construction and was also deposited in the RIKEN BioResource Research Center (BRC), Tsukuba, Japan (BRC catalog no. RDB14361).

2.3 Construction of *adenylyl cyclase A* reporter vector

The *Dictyostelium* extrachromosomal expression vectors pDM304 (NBRP ID: G90008) and pDM358 (NBRP ID: G90009) were provided by Tsukuba

University under the NBRP and were used in the construction of two *acaA* reporter vectors involving promoters 1 and 3. The actin 15 promoter (*Xho*I/*Bgl*II restriction sites) of the pDM358 and pDM304 vectors was replaced with promoters 1 and 3 of *acaA*, respectively. The promoter regions were amplified using PCR from *Dictyostelium discoideum* Ax2 genomic DNA using the primer sets for promoter 1 (5′-GCctcgagCTTGATGAGTGGCCCAAAACC-3′ and 5′-GCagatctATTTTTAAAGATCCAAGAATTCG-3′) and promoter 3 (5′-GCctcgagACCTCACTTCATAAATATATCTTTG-3′ and 5′-GCagatct TTTTTAATAATTTTTTAATATTATTAC-3′) [10]. Variant 1 of *Luci sp1* (green) and *Psa* (red) luciferase genes, including the *Dictyostelium* Kozak sequence (AAAA) before the start codon, was inserted into downstream of the promoter in the pDM358 and pDM304 vectors, respectively, *via* the *Spe*I/*Hind*III restriction sites. These vectors contained the actin 8 terminator.

2.4 Transformation and fruiting body formation

The Ax2 cells were co-transfected with the two constructed vectors by electroporation using a MicroPulser (Bio-Rad, California, USA) according to the protocol for *Dictyostelium*. The transformant cells were selected using hygromycin (50 mg/mL) and neomycin (10 mg/mL) in HL5 medium.

Millicell Cell Culture Insert PICM 0RG50 (Merck, Darmstadt, Germany) was placed onto a 35-mm glass bottom dish, and 1.2 mL of D buffer (saline solution for *Dictyostelium*) containing 3 mM D-luciferin potassium salt (Promega, Wisconsin, USA) was added to the basolateral side of the glass bottom dish. The transformant cells were seeded onto the inside of the cell culture insert above the membrane and cultured at room temperature (21°C), facilitating the development of the mound, slug, and fruiting body.

2.5 Bioluminescence microscopy

The bioluminescence images of the cells were captured using an LV200 bioluminescence microscope (Olympus, Tokyo, Japan) [19, 20] equipped with a UCPLFLN 20XPH objective lens (Olympus) and an ImagEM C9100-13 EM-CCD camera (Hamamatsu Photonics, Shizuoka, Japan). The activities of *acaA* promoters 1 and 3 were visualized using BA495-540GFP (Olympus) and 610ALP (Omega, New Jersey, USA) emission filters, respectively, at an exposure time of 30 s for 24 h at 90s intervals. In addition, bright-field images were captured at an exposure time of 200 ms using the same capture sequence as that of the bioluminescence images. The time course of luminescence intensity was analyzed using TiLIA, a time-lapse image analysis software [21].

2.6 Fluorescence microscopy

The autofluorescence images of the cells were captured using an IX83 inverted microscope (Olympus) equipped with a U-HGLGPS excitation light source (output level 100 with ND25 filter), a U-FGFP mirror unit, and a DP74 color CCD camera. A UCPFLN 10XPH objective lens was used for mound and slug observations, and a UCPFLN 4xPH objective lens was used for fruiting body observation. Exposure time was 500 ms for all experiments.

The irradiation power of the excitation light was measured at 480 nm using a PM100D optical power meter (Thorlabs, New Jersey, USA) with an S170C sensor probe for microscopy.

3. Results and discussion

Bright-field and bioluminescence images of the activities of promoters 1 and 3 are shown in **Figure 1**. According to the bright-field image, the amoeba cells began to aggregate after 10 h of seeding (**Figure 1A**) and formed the mound (**Figure 1D**), slug, and fruiting body (**Figure 1F**) after 16, 18, and 20 h, respectively. The image of the slug is not shown in **Figure 1**, since the moving slug disappeared from the field of view. In the case of **Figure 1**, the slug stage was extremely short and the fruiting body formation occurred immediately from the mound.

According to the bioluminescence image, promoter 1 activity was observed in single amoeba cells. It increased gradually (**Figure 1A–C**) and peaked at the mound

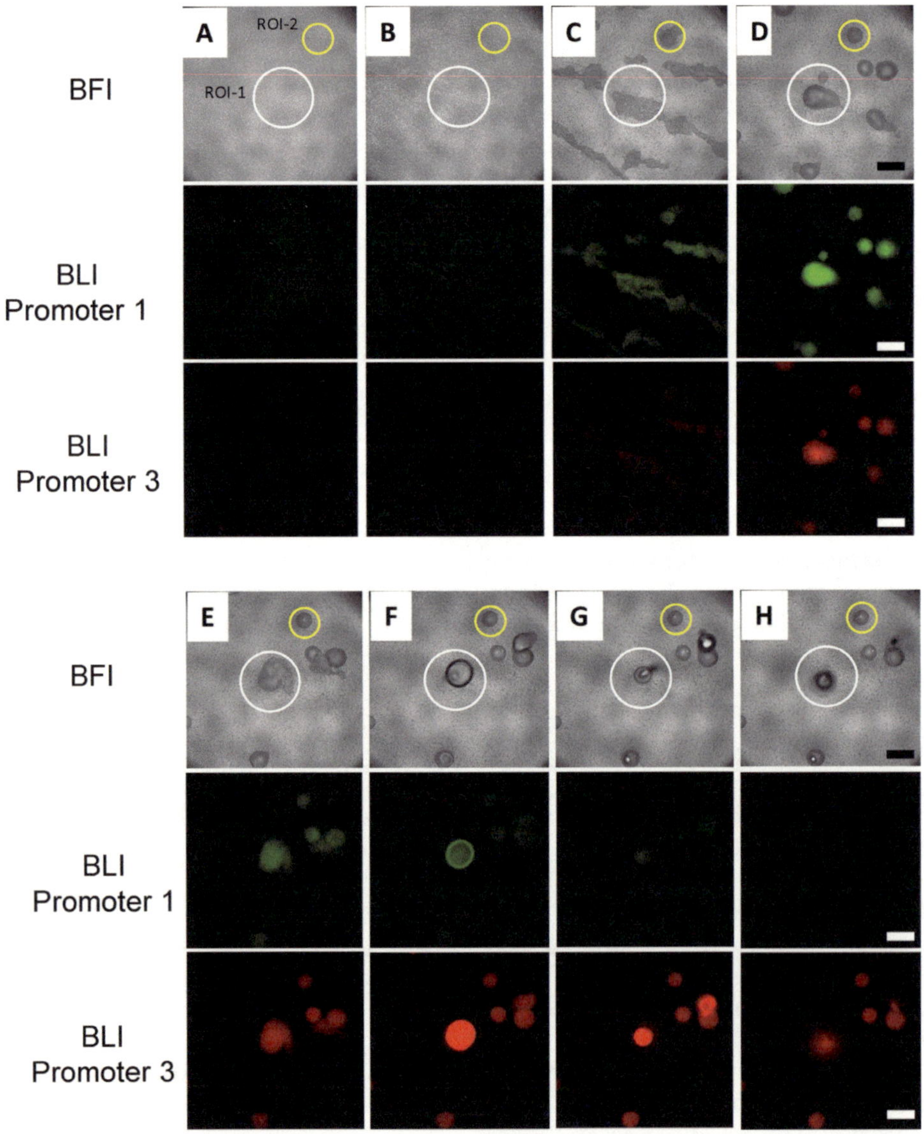

Figure 1.
Bright-field images (BFI) and bioluminescence images (BLI) reflecting the activities of promoters 1 and 3 at 10 h (A), 12 h (B), 14 h (C), 16 h (D), 18 h (E), 20 h (F), 22 h (G), and 24 h (H) after seeding ameba cells. Regions of interest 1 and 2 (ROI-1 and ROI-2) were assigned to cover the process from cell aggregation to fruiting body formation and are marked as circles on the BFI. Scale bar: 500 mm.

stage after 16 h (**Figure 1D**). Then, it decreased during fruiting body formation (**Figure 1E–G**) and eventually disappeared (**Figure 1H**). On the other hand, promoter 3 activity increased during the cell aggregation stage (**Figure 1C**), peaked during the fruiting body stage (**Figure 1F**), and then decreased (**Figure 1G** and **H**). Thereafter, the activity of the stalk cells disappeared. Histochemical detection of promoter activity using a *lacZ*/X-Gal staining system [10] showed that promoter 1 activity was detected in the cell aggregation and early mound stages, but was not detected after the slug stage. On the other hand, promoter 3 activity was detected in the early mound, slug, and spore stages. This activity was particularly strong in the upper cup of the spores. The two methods yielded almost the same results, but

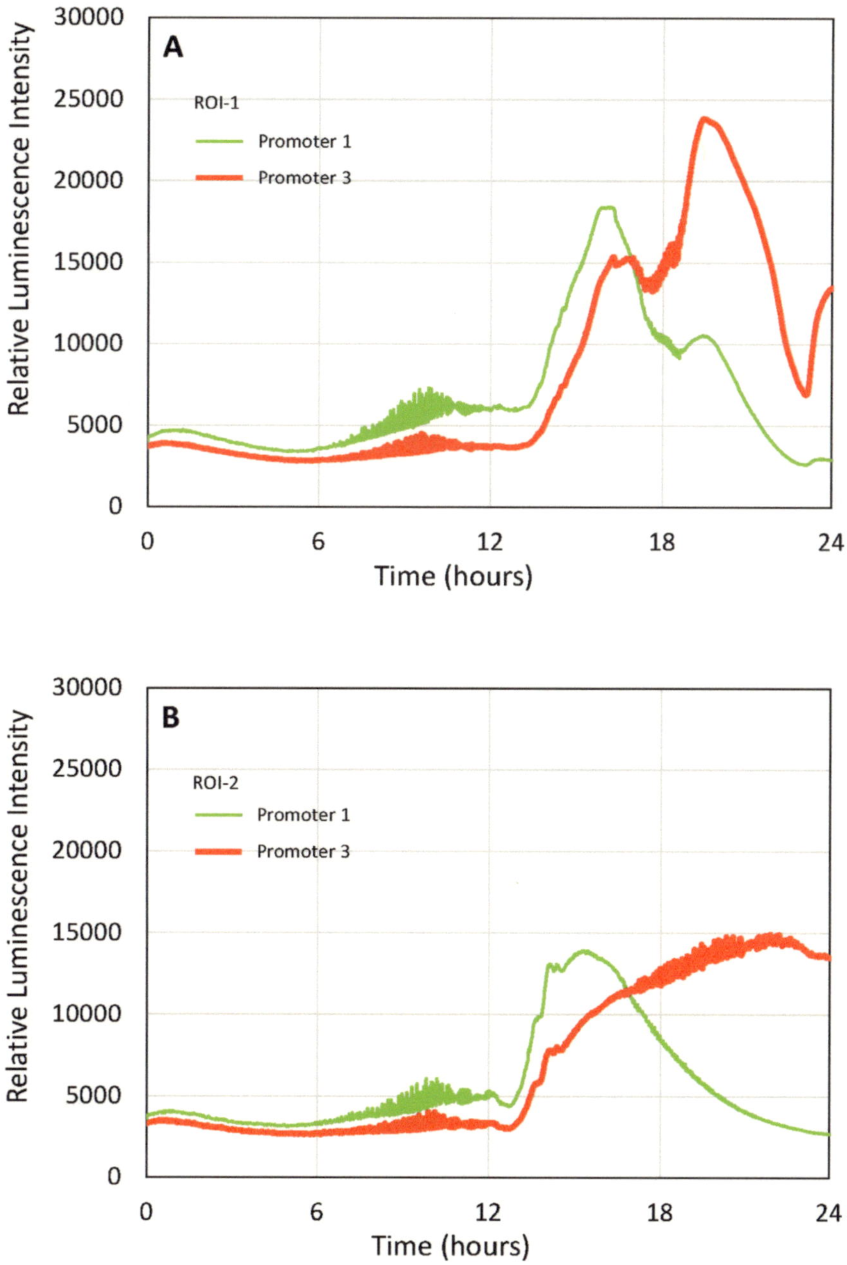

Figure 2.
Time course of luminescence intensity reflecting the activities of promoters 1 and 3 in ROI-1 (A) and ROI-2 (B).

promoter 3 activity was detected earlier by bioluminescence imaging than by histochemical detection at the cell aggregation stage after 14 h (**Figure 1C**). However, the resolution of the images obtained by histochemical detection was superior to that by bioluminescence imaging.

To show the time course of the promoter activities, two regions of interest (ROI) (ROI-1 and ROI-2) were assigned to cover the process from cell aggregation to fruiting body formation, as shown in **Figure 1**. **Figure 2** shows the time course of luminescence intensity reflecting the activities of promoters 1 and 3 in ROI-1 (**Figure 2A**) and ROI-2 (**Figure 2B**) using 961 time-lapse images captured at 90s intervals for 24 h. The intensity of promoter 1 activity increased after 13 h, peaked after 16 h, decreased, and disappeared after 24 h in ROI-1 and ROI-2. On the other hand, the intensity of promoter 3 activity increased to the same timing as that of promoter 1 after 13 h, but peaked after around 20–22 h. Then, the intensity decreased gradually in ROI-2, but rapidly decreased and recovered in ROI-1. Since the measurement of the intensities involves live imaging, the discrepancy may be caused by the movement of the spores in the ROI during fruiting body formation. The measurement of the time courses of the activities of promoters 1 and 3 by bioluminescence imaging and by a β-galactosidase reporter system [10] showed similar results.

The results of the promoter assay using bioluminescence microscopy were the same as those of the promoter assays using histochemistry and β-galactosidase, confirming the convenience of this imaging promoter assay for *Dictyostelium* studies. Moreover, the imaging promoter assay enabled the spatiotemporal information of promoter activity to be obtained sequentially in a single experiment. Based on the result, detailed analysis of promoter activity can be performed efficiently by histochemical or immunofluorescence microscopy. However, several experiments are required for each measurement in histochemical and β-galactosidase promoter assays. Multiple promoters can also be analyzed using multicolor luciferases, but the number of promoters evaluated is limited to the number of luciferases of different colors. Moreover, there are some concerns regarding the use of multicolor luciferases in bioluminescence microscopy as follows. **Figure 3** shows the normalized

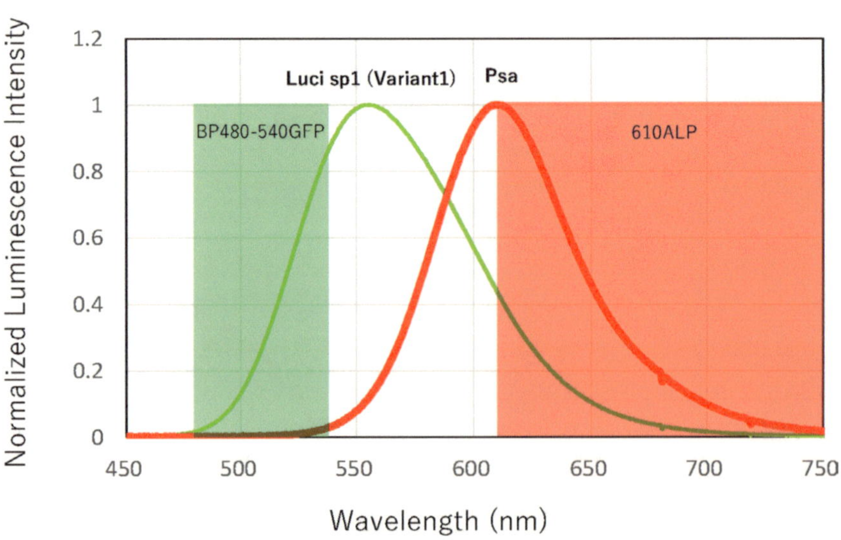

Figure 3.
Normalized luminescence spectra of luciferases (variant 1 of Luci sp1 and Psa) expressed in HeLa cells in the transparent range of the emission filters, BP480-540GFP and 610ALP.

luminescence spectra of the luciferases (variant 1 of *Luci sp1* and *Psa*) in the transparent range of the emission filter for each luciferase channel. The luciferases were expressed in HeLa cells [17, 18]. In the transparent range of the 610ALP filter, cross talk between the two spectra was observed between 610 and 700 nm. To prevent spectral cross talk, a spectral unmixing operation must be done, as is performed in fluorescence microscopy [22]. Nakajima et al. [23] demonstrated the unmixing of tri-colored bioluminescence for a luciferase promoter assay using one color to normalize the activity of two genes. In addition to the number of promoters to evaluate, we need one more color luciferase to normalize different promoter activities.

One of the advantages of bioluminescence microscopy is that it is not affected by autofluorescence background. **Figure 4** shows the bright-field and autofluorescence images of the mound, slug, and fruiting body stages of *Dictyostelium discoideum* development captured by fluorescence microscopy with a mirror unit for green fluorescent protein (GFP). Autofluorescence from the upper tip of the mound, the periphery of the slug body, and the spore and stalk of the fruiting body was observed. Therefore, the imaging conditions used GFP as a reporter require optimization of the excitation intensity, etc.

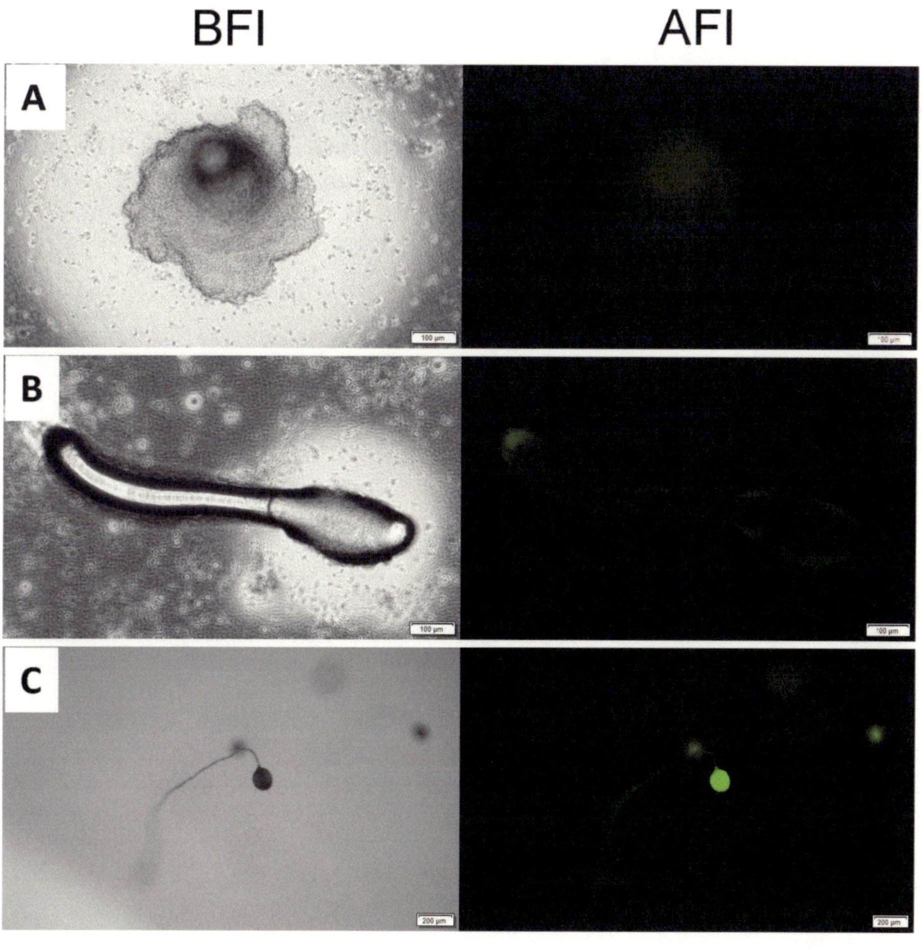

Figure 4.
Bright-field images (BFI) and autofluorescence images (AFI) of the mound (A), slug (B), and fruiting body (C) stages of Dictyostelium discoideum *development. Exposure time was 500 ms, and excitation light power was 0.8 mW for the mound and slug stages and 1.0 mW for the fruiting body stage. Scale bar: 100 mm for A and B and 200 mm for C.*

4. Conclusion

The imaging promoter assay of the *acaA* promoters 1 and 3 by bioluminescence microscopy and the histochemical and β-galactosidase promoter assays yielded similar results in the evaluation *acaA* promoter activity in the different stages of *Dictyostelium discoideum* development. Moreover, we found that obtaining an overall picture of promoter activity spatiotemporally during the entire developmental process is possible with the bioluminescence imaging promoter assay. However, spectral unmixing is required to effectively normalize the different promoter activities. We also found that the imaging promoter assay is not significantly affected by autofluorescence.

Acknowledgements

We thank Prof. H. Kuwayama (Tsukuba University, Japan) for providing the materials needed in the *Dictyostelium discoideum* experiments, offering valuable advice, and allotting time for discussion. We also thank our colleagues at Olympus Corporation for providing technical assistance during the development of the LV200 bioluminescence microscope.

This work was conducted in association with Nimura Genetic Solutions and Perak State Development Corporation for luciferase gene cloning under a contract with the Convention on Biological Diversity in Malaysia. This research did not receive grants from funding agencies in the public, commercial, or non-profit sectors.

Conflict of interest

The authors T. Hayashi, K. Ogoh, and H. Suzuki are employees of Olympus Corporation (Tokyo, Japan).

Author details

Taro Hayashi[1*], Katsunori Ogoh[1] and Hirobumi Suzuki[1,2]

1 Olympus Corporation, Tokyo, Japan

2 Department of Applied Chemistry, School of Advanced Engineering, Kogakuin University, Tokyo, Japan

*Address all correspondence to: taro.hayashi@olympus.com

References

[1] Konijn TM, van de Meene JGC, Bonner JT, Barkley DS. The acrasin activity of adenosine-3′,5′-cyclic phosphate. Proceedings of the National Academy of Sciences of the United States of America. 1967;**58**:1152-1154

[2] Dormann D, Vasiev B, Weijer CJ. The control of chemotactic cell movement during *Dictyostelium* morphogenesis. Philosophical Transactions of the Royal Society of London B. 2000;**355**:983-991

[3] Barklis E, Lodish HF. Regulation of *Dictyostelium discoideum* mRNAs specific for prespore or prestalk cells. Cell. 1983;**32**:1139-1148

[4] Mehdy MC, Ratner D, Firtel RA. Induction and modulation of cell-type-specific gene expression in *Dictyostelium*. Cell. 1983;**32**:763-771

[5] Hopper NA, Anjard C, Reymond CD, Williams JG. Induction of terminal differentiation of *Dictyostelium* by cAMP-dependent protein kinase and opposing effects of intracellular and extracellular cAMP on stalk cell differentiation. Development. 1993; **119**:147-154

[6] Saran S, Meima ME, Alvarez-Curto E, Weening KE, Rozen DE, Schaap P. cAMP signaling in *Dictyostelium*. Complexity of cAMP synthesis, degradation and detection. Journal of Muscle Research and Cell Motility. 2002;**23**:793-802

[7] Kriebel PW, Parent CA. Adenylyl cyclase expression and regulation during the differentiation of *Dictyostelium discoideum*. IUBMB Life. 2004;**56**:541-546

[8] Pitt GS, Milona N, Borleis J, Lin KC, Reed RR, et al. Structurally distinct and stage-specific adenylyl cyclase genes play different roles in *Dictyostelium* development. Cell. 1992;**69**:305-315

[9] Verkerke-van Wijk I, Fukuzawa M, Devreotes PN, Schaap P. Adenylyl cyclase A expression is tip-specific in *Dictyostelium* slugs and directs Stata nuclear translocation and CudA gene expression. Developmental Biology. 2001;**234**:151-160

[10] Galardi-Castilla M, Garciandía A, Suarez T, Sastre L. The *Dictyostelium discoideum* acaA gene is transcribed from alternative promoters during aggregation and multicellular development. PLoS One. 2010;**5**:e13286

[11] Ukai H, Kobayashi TJ, Nagano M, Masumoto KH, Sujino M, Kondo T, et al. Melanopsin-dependent photo-perturbation reveals desynchronization underlying the singularity of mammalian circadian clocks. Nature Cell Biology. 2007;**9**:1327-1334

[12] Yagita K, Horie K, Koinuma S, Nakamura W, Yamanaka I, Urasaki A, et al. Development of the circadian oscillator during differentiation of mouse embryonic stem cells in vitro. Proceedings of the National Academy of Sciences of the United States of America. 2010;**107**:3846-3851

[13] Kwon H, Enomoto T, Shimogawara M, Yasuda K, Nakajima Y, Ohmiya Y. Bioluminescence imaging of dual gene expression at the single-cell level. BioTechniques. 2010;**48**:460-462

[14] Akiyoshi R, Kaneuch T, Aigaki T, Suzuki H. Bioluminescence imaging to track real-time armadillo promoter activity in live drosophila embryos. Analytical and Bioanalytical Chemistry. 2014;**406**:5703-5713

[15] Morishita Y, Kuroiwa A, Suzuki T. Quantitative analysis of tissue deformation dynamics reveals three characteristic growth modes and globally aligned anisotropic tissue deformation during chick limb

development. Development. 2015; **142**:1672-1683

[16] Akieda Y, Ogamino S, Furuie H, Ishitani S, Akiyoshi R, Nogami J, et al. Cell competition corrects noisy Wnt morphogen gradients to achieve robust patterning in the zebrafish embryo. Nature Communications. 2019;**10**:4710

[17] Ogo K, Akiyoshi R, Suzuki H. Firefly luciferase. United States Patent 2014; US8722376B2

[18] Ogoh K, Akiyoshi R, Suzuki H. Cloning and mutagenetic modification of the firefly luciferase gene and its use for bioluminescence microscopy of engrailed expression during *Drosophila* metamorphosis. Biochemistry and Biophysics Reports. 2020;**23**:100771

[19] Ogoh K, Akiyoshi R, May-Maw-Thet, Sugiyama T, Dosaka S, Hatta-Ohashi Y, et al. Bioluminescence microscopy using a short focal-length imaging lens. Journal of Microscopy. 2014;**253**:191-197

[20] Suzuki H, May-Maw-Thet, Hatta-Ohashi Y, Akiyoshi R, Hayashi T. Bioluminescence microscopy: Design and applications. In: Thirumalai J, editor. Luminescence—An Outlook on the Phenomena and Their Applications. Rijeka, Croatia: Intech; 2016. pp. 333-349

[21] Konno J, Hatta-Ohashi Y, Akiyoshi R, Thancharoen A, Silalom S, Sakchoowong W, et al. TiLIA: A software package for image analysis of firefly flash patterns. Ecology and Evolution. 2016;**6**:3026-3031

[22] McRae TD, Oleksyn D, Miller J, Gao YR. Robust blind spectral unmixing for fluorescence microscopy using unsupervised learning. PLoS One. 2019;**14**:e0225410

[23] Nakajima Y, Kimura T, Sugata K, Enomoto T, Asakawa A, Kubota H, et al. Multicolor luciferase assay system: One-step monitoring of multiple gene expressions with a single substrate. BioTechniques. 2005;**38**:891-894

Section 2

Bioluminescent Organisms

Chapter 5

Firefly Translocation: A Case Study of Genetic and Behavioral Evaluation in Thailand

Anchana Thancharoen

Abstract

Conservation translocation is frequently used to conserve the threatened fauna by releasing individuals from the wild or captive populations into a particular area. This approach, however, is not successful in many cases because the translocated populations could not self-sustain in the new habitats. In this chapter, I reviewed the concept of translocation for conservation and the factors associated with the success rate. I used example problems from several cases involving different insect taxa. With its often high potential to mass rear in captivity, captive breeding can be a powerful tool by assuring large population size for insect translocation, which can result in a high success rate. However, genetic consequences from inbreeding and genetic adaptation to captivity can reduce the fitness of the captive population to establish successfully in the wild. Additionally, as the evidence in Japanese fireflies shows, the genetic differences between the translocated and local populations should be considered for a sustainable translocation program. A case study involved genetic and behavioral evaluation of *S. aquatilis* populations to assess the possibility of including the species for the firefly translocation program in Thailand. Although the results revealed no genetic variation among populations, examination of the variation in flash signals showed that the long-distance population had a longer courtship flash pulse than other populations in the Bangkok Metropolitan Region. With no geographical barrier, the light pollution and urbanization are probably important fragmented barriers causing adaptation of flash communication to increase the fitness. As a consequence, firefly translocation should consider flash variation between populations to prevent this potential pre-mating isolation mechanism from resulting in probable lower translocation success rates.

Keywords: Lampyridae, aquatic firefly, *Sclerotia aquatilis*, flashing behavior, population genetic, intraspecific variation, TiLIA software

1. Introduction

Fireflies have long been attracted the attention of people because of their fascinating flashing communication behavior [1]. In the past, firefly flashes on mangrove trees along the river were used as landmarks for boat navigation in the nighttime; while nowadays firefly habitats become "firefly tour sites" for nighttime activity and for supporting economic benefit to local communities [2]. Unfortunately, firefly populations decrease or disappear from many areas worldwide due to habitat loss from growing of city developments, light pollution, water pollution and pesticide uses, which cause habitat destruction or fragmentation [3–7]. This same situation is faced by other insects [8]. In addition, firefly tourism without proper management could result in decreased firefly populations [2, 9, 10]. The problem has, thus, led to increased public awareness of firefly conservation.

Firefly conservation by reintroduced captive populations into the wild has received much attention. The successful captive breeding of some firefly species has intrigued numerous naturalists and conservationists including tourism stakeholders to plan to introduce captive breeding firefly populations into many areas to create firefly conservation sites, environmental learning centers and firefly tourism spots. The firefly mass rearing has been successful in some aquatic species, including *Aquatica leii* [11], *A. ficta* [12], *A. hydrophila* [13], *A. lateralis* [14], *S. aquatilis* [15, 16], and *S. substriata* [17]. A few of them have been used for conservation translocation. Many parks in Taipei, Taiwan were restored for suitable habitat and captive bred *A. ficta* fireflies were released [18–22]. In Korea, *L. lateralis* habitat (both running water and lentic water areas) was artificially created for releasing the mass reared populations of the species for ecotourism purposes [23]. As a symbol of nature in Japan, many firefly reintroduction and restoration projects of *L. cruciata* and *L. lateralis* have been done over the centuries, but not all of them have been successful [24]. Unfortunately, there are many cases showing strong ecological impact of introduced firefly populations on the native populations, which might eventually lead to the loss of the native populations in Japan [25]. This problem occurs where there is geographical isolation, based on examined differences of flash rate and genetic studies [26]. Therefore, the study of the impact of firefly translocation is essential prior to implementation of the program. Such impact studies have been lacking in Thai firefly translocation projects. Background information on genetic and behavioral variations among populations is necessary for development of a sustainable firefly reintroduction programs.

2. General aspects of translocations for conservation

Conservation translocation (population restoration) or called "ex situ conservation." Under the definition of the IUCN this is the intentional movement of released organisms from one to another site for conservation benefits [27]. That consists of two terms: (i) "reinforcement" which is augmenting a species where it already exists and (ii) "reintroduction" which is returning a species back to where it has disappeared [28]. With the increasing of habitat loss and fragmentation resulting in high species extinction rates and reduction of overall biodiversity, translocation of species may become an important management tool for recovery of the diminished or lost populations.

Many translocation programs have been carried out in many rare, threatened and keystone species to conserve species and genetic diversity. For example, European bison [29], Lake Sturgeon [30], Persian wild ass [31], green and golden bell frog [32], red wolves [33], and a few insects, (i.e., damselfly [34], field cricket [35] and fireflies [25]). Most of them have involved vertebrates, especially mammals and birds [36]. Consequently, translocation became an important conservation technique for birds in New Zealand [37]. However, as mentioned above, little work has been done in insect taxa.

The success of translocations was defined as resulting in self-sustaining populations in the release area. The success rate is affected by many factors. For example, species, habitat quality of the release areas, location of the release point, origin of animals (captivity or wild), food habit (carnivore, herbivore and omnivore), clutch size, population density and competitors [36]. The research analyzed from translocation studies of 134 bird and 64 mammal projects concluded that the keys for high translocation success rate were releasing wild-caught animals, having herbivore food habits, releasing a large density, releasing in excellent quality habitats and releasing at the center of the area. In addition, the reproduction rate and generation length might affect the population sizes, chances of survival and genetic diversity of the target [38].

Many problems of population establishment from translocation were investigated. The small released populations might result in demographic and genetic consequences, for example, inbreeding depression [38]. Moreover, in the cases of releasing of a captive breeding population, the captive-born individuals provided from benign and stable breeding environments frequently have reduced fitness and high extinction rates after release into the wild. The physiological, behavioral and ecological problems from inbreeding depression, mutation accumulation, loss of genetic diversity and genetic adaptation to captivity were considered [39–43]. These could affect success of translocation programs through low adaptive potential to environmental changes [44]. Thus, many recommendations for dealing with the genetic issues are as follow: (i) minimizing numbers of generations in captivity, (ii) maintaining isolated captive populations with different genetic strains to reduce genetic load, (iii) allowing half-sib mating in captivity to reduce genetic adaptation to captivity and preserve genetic variation, (iv) minimizing kinship by equalizing family sizes and crossing, (v) observing the behaviors that might be lost in captivity, (vi) creating a rearing environment similar to the natural habitat to minimize the artificial selection, (vii) evaluating other risks (i.e., diseases), (viii) and collecting and analyzing long-term monitoring data routinely [39, 41–42, 45–47]. Although returning a lost species might not be same as the outcome of ecosystem restoration, the species perform ecosystem functions and generally relate to the other species. Polak and Saltz [48] suggested that the study on the effects of reintroductions on ecosystem functions should be integrated into the programs. Further, an overlooked issue of genetic impact is genetic contamination by maladaptive genotypes from reproductive crossing between genetically differentiated populations. That could push the recipient population toward extinction [49]. Therefore, the introgression with the population having local genetic makeup could result in a well-adapted population with similar morphological and ecological characters to local types.

3. Translocations in insects

The translocation of insects and other invertebrates has received considerably less attention than vertebrates; thus, not many examples of insects were translocated. However, ex situ conservation has become recognized as a more important technique for conservation for insects. With small body size, high reproductive rates, and short generation times, the insects have high potential to breed in mass captivity involving lower maintenance costs. Pearce-Kelly et al. suggested that the easy-breeding species with large captive populations have high potential for successful reintroduction programs [50]. The summary of 134 terrestrial insect translocations demonstrated that the proportion of success (52%) was higher than other animals while failed translocation programs were lower, 31% [51]. Thus, insects are the group most frequently considered in future translocations [52].

The objectives of insect translocation were classified into two groups, for conservation of the rare species and for socio-economic benefits of the flagship species.

Examples of the rare insect translocation are two vulnerable crickets, *Gryllus campestris* and *Decticus verrucivorus*, in England [53–54], the threatened tiger beetle *Cicindela dorsalis dorsalis* [55], a rare damselfly *Ischnura gemina* [56], Quino checkerspot butterfly *Euphydryas editha quino* [57] and the Genji firefly *Luciola cruciata* [58] (**Table 1**). With several iterations of releasing, the released insects could establish over a period of time and produced subsequent self-sustaining populations. The failure of translocation cases were caused by small released populations, disease infection, high dispersal stage used for releasing, low quality of habitat and weather conditions when releasing. The previous study [59] analyzed the documentations of 50 reintroduction activities of butterfly species and concluded that the successful projects had a higher number of attempts (per species) (11.1 ± 11.3 times for successful and 3.5 ± 3.2 times for unsuccessful programs). Successful programs introduced at least 292 individuals per reintroduction and continued for three years. Significantly, captive breeding was recommended for reintroduction programs for almost 50% of butterfly species.

As a dominant invertebrate flagship, the translocation of butterflies could be effectively used to build public awareness using live exhibits of butterfly farms. Many exotic butterflies were large-scale bred and imported across countries and regions for exhibition. If the butterflies come from similar environmental conditions and habitats, they might have high potential to establish in the new habitats. Consequently, the unintentional translocation might happen and cause ecological

Insects	Threats	Sources of translocated population	Success?	Problems of the translocation
Field cricket *G. campestris*	Rare and fragmented habitats	Captivity	Success (5 years)	- disease infection - cannibalism
Wart-biter bush cricket *D. verrucivorus*	Rare and fragmented habitats	Captivity	Failure	- high mortality rate in captivity result in small translocated population - high rearing cost
Tiger beetle *C. dorsalis dorsalis*	Sandy beach habitats of larvae were destructed from increasing of recreational activity.	Field collection (larvae)	Success (8 years)	- failure in adult translocation because of high dispersal behavior - larval predation by gulls
Damselfly *I. gemina*	Habitat structure changes and water area destruction from urbanization	Field collection (mating pairs)	Success (1 year in beginning phases)	- habitat changes from over vegetation in 2nd year. - unsuitable handling and marking techniques
Quino checkerspot butterfly *E. editha quino*	Habitat loss, fragmentation and extinction of native host plants	Captivity	Success	N/A
Genji firefly *L. cruciata*	Habitat loss, water pollution and tourism activities	Field collection and captivity	Success (70 years)	- harvested high amount of fireflies and released the non-native populations

Table 1.
Comparison of factors in some examples of rare insect translocation programs.

impact [60]. The opposite effect also may result, that captive bred populations lose the ability to live in natural habitats. After breeding in captivity for 100–150 generations, the large white butterfly have developed adaptive characters to captive conditions, i.e., heavier, higher ovary mass, higher numbers of laid eggs, and smaller wings that could decrease the butterflies' ability to re-establish in the wild [61].

The firefly is also a potential flagship to stimulate conservation awareness and action to support habitats for fireflies and other sympatric invertebrates. Apparently, firefly populations have declined or become extinct in many areas due to the impact of anthropogenic activities (i.e., habitat destruction, fragmentation, pollution and urbanization). Fireflies can be used to help promote public awareness and concern for biological diversity conservation.

The history of firefly translocation probably began in Japan [58]. The famous case happened in Tatsuno, Nagano prefecture where several thousand of the non-native Genji fireflies from Shiga prefecture were released as a tourist attraction. Subsequently the variation in flashing behavior and population genetics were investigated. Although the population of Genji fireflies in Tatsuno could self-establish over 70 years in the translocated area and bring more than 100,000 tourists a year, the native populations might be destroyed or lose genetic diversity. That is the risk under environmental change in the upcoming global crisis. Later, the scientists raised awareness of the firefly conservation issue and recommended the approach of using habitat preservation instead of artificial habitat creation for tourism. The fireflies were commonly labeled as an indicator species for environmental conservation. The translocation of captive fireflies in recovering polluted environments received more attention and resulted in appearance of 540 firefly events throughout Japan.

4. Genetic variation among firefly populations: the difficulty in translocation

Genetic issues become more important in sustainable biodiversity conservation especially in animal translocation. Avoiding or reducing genetic problems is a key to reducing the risk of extinction. Thus, not only focusing on maximizing species survival from established population measures, but also focusing on the genetic diversity, genetic drift and genetic adaptation to captivity are necessary to evaluate viability of populations in the long term.

The evidences of genetic and behavioral variation among firefly populations in Japan were discussed above. Firefly translocation requires an appropriate evaluation prior to their introduction into the wild. Likewise, the long term post-monitoring of both genetic and phenotypic measures is needed to measure the success of translocation and to identify future threats.

Genetic differentiation of fireflies is caused by various factors, including limitation of dispersal activity, habitat specificity or mating systems. The species with limited dispersal species have a higher probability of reproductive isolation. As in the desert firefly *Microphotus octarthrus*, which have winged males and apterous larviform females, the discontinuous habitats results in genetic isolation [62]. Strong habitat specificity was apparently involved, and there are several other cases of genetic divergence of fireflies influenced by geographical isolation. The variation of genetic structure of *Pyrocoelia rufa* in Korea was examined among islands, western and earthen parts being separated by mountain barriers resulting in different habitat types [63]. Consistently, the variation of genetic and phenotypic patterns of several firefly species in Japan was geographically separated by the Itoigawa-Shizuoka tectonic line. *Hotaria parvula* with morphological variation of body size are associated with genetic differentiation and are reproductively isolated [64]. Likewise, two population groups of *L. cruciata* in eastern and western areas of the tectonic line were also genetically different and displayed different flash communication patterns (slow-flash and fast-flash types) [65]. The variation in male flash patterns (based on inter-flash interval) was subsequently confirmed to have the potential to hinder in pre-mating between populations. The intermediate flash type fireflies that might be introgressive hybridization were found near the barrier area [66, 67]. Surprisingly, the "quick-flash type" was investigated in the Goto islands, the western tip of Kyushu but it was in the same haplotype as the fast flash fireflies inhabiting the mainland [68]. On the other hand, *A. lateralis* populations throughout the Korean Peninsula, northeast China, Sakhalin, and Japan were examined for genetic variation of two flash pattern types (which also have a difference in adult emergence season duration) but they could not be separated phylogenetically [69].

5. A case study of genetic and behavioral evaluation of Thai firefly species, *Sclerotia aquatilis*

5.1 Background

Sclerotia aquatilis (*L. aquatilis*) [70] is an aquatic firefly species. Individuals are commonly found in freshwater habitats throughout Thailand, i.e., ponds, ditches, wetlands inhabited by an abundance of aquatic snails and aquatic vegetation such as duck weed, water lettuce, water hyacinth, *Typha* spp., water lily, and Indian lotus. It is a multivotine species appearing all year round with the life cycle duration of 3–5 months [71], **Figure 1**. The larvae live in the water by respiring mainly through a pair of caudal spiracles to receive the air from water surface. They are frequently found back swimming at the surface of water.

The species has high potential for reintroduction programs because of the successful rearing technique developed [15, 16] and their several adaptive characteristics that support recovery of the new populations in old/new habitats. Since *S. aquatilis* occurs throughout Thailand, the reintroduction programs are probably

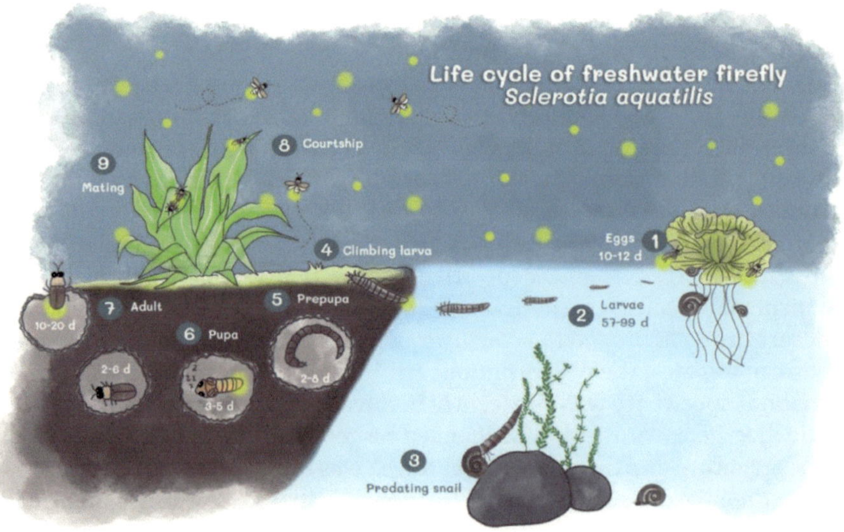

Figure 1.
Life cycle of S. aquatilis.

applied widely in the country. The firefly translocation has not previously been reported for this species.

There are many reasons suggesting genetic differentiation among *S. aquatilis* populations could lead to negative impact on translocation programs. Although geographic isolation frequently results in reproductive isolation by limiting gene flow between populations, it still remains unknown for firefly populations in Thailand. The expansion of cities and associated infrastructures not only destroy firefly habitats, but also creates habitat fragmentation. *S. aquatilis* populations are restricted to freshwater habitats, i.e., ponds, wetlands, and ditches. Adult female fireflies lack strong flight ability; therefore, habitat fragmentation seriously limits the range of their dispersal efforts, resulting in little immigration and even local extinctions. These limiting dispersal factors cause an increased the level of inbreeding and minimize interbreeding among spatially isolated populations. Thus, the probability of inbreeding and low genetic variability in nature is high in fragmented habitats. There is evidence of loss of genetic variation and the extinction of populations from habitat fragmentation in a butterfly metapopulation [72]. In addition, most *S. aquatilis* habitats overlap with human-used areas such as residential and agricultural areas, fireflies are subjected to many negative impacts from human urbanization, especially light pollution that can interfere with the sexual communication signals. Moreover, light pollution can be an effective dispersal barrier of fireflies. All these factors might result in both decreasing numbers and promoting inbreeding effects in populations.

5.2 Materials and methods

5.2.1 Study areas

During the process of urbanization, habitat loss and fragmentation have subsequently expanded particularly in Bangkok (BKK) area, where is the focus area for firefly reintroduction in this study. Historically, *S. aquatilis* inhabited in high abundance in the agricultural diches and ponds in the Chao Phraya delta area. However, the recent populations of the species have been decreased and become rare. The sources of translocated populations were from four nearby provinces, Samut Prakarn (SPK), Pathum Thani (PTE), Nakhon Pathom (NPT), and Suphan Buri (SPB) (**Figures 2** and **3**). Seven populations of fireflies from five locations were collected. One population from each province but two subpopulations from Pathum Thani (PTE2) and Nakhon Pathom (NPT2).

5.2.2 Firefly collection and maintenance

The collection of *S. aquatilis* specimens was conducted in all five locations during firefly season from August to November in 2012–2013, which was during the end of the raining season and the beginning of winter. The adult fireflies were collected at nighttime using a sweep net over freshwater areas. Adults were maintained in insect rearing cages supplied with a 10% honey solution on balls of moist cotton. In case of small populations, aquatic firefly larvae were also collected for molecular work. After observing the flashing behavior, the firefly specimens were placed in vials containing 100% ethanol, and stored in a − 80°C freezer prior the molecular study.

5.2.3 Genetic analysis

Genomic DNA from the hind legs of the adult specimens was extracted following the manufacturer's protocol using the DNeasy Blood & Tissue

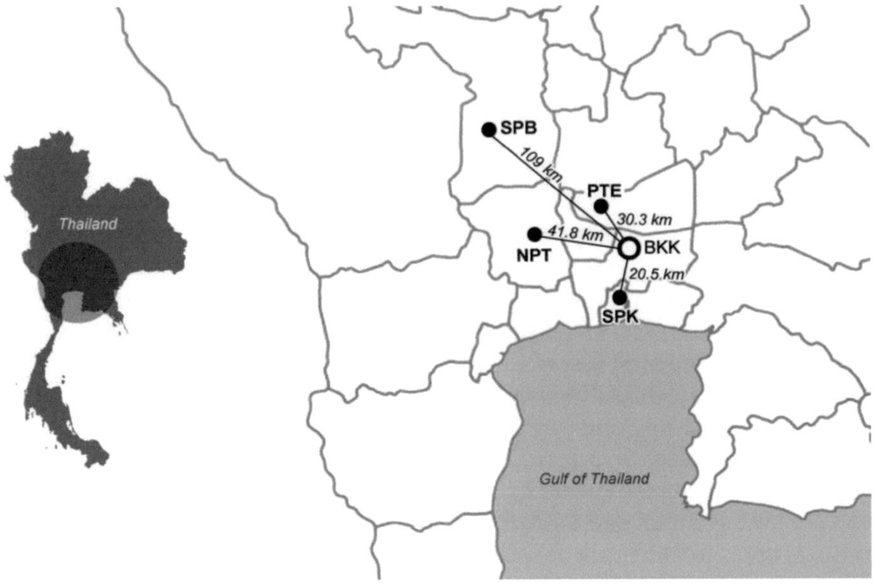

Figure 2.
Map of Thailand the S. aquatilis study sites. The map illustration was modified from Vemaps.com.

Kit (Qiagen). A region encoding mitochondrial cytochrome c oxidase subunit II (COII) was amplified by the polymerase chain reaction (PCR) using the primers 5′-ATGGCAGATTAGTGCAATGG-3′ (TL2-J-3037) and 5′-GTTTAAGAGACCAGTACTTG-3′ (TK-N-3785) [69]. The PCR amplifications were performed as follows: an initial denaturing step at 94°C for 1 min, followed by 35 cycles beginning with a denaturation step at 94°C for 30 sec, an annealing step at 50°C for 30 sec, an extension step at 72°C for 1 minute, and a final step at 72°C for 10 min. The PCR product was verified by running through a 1% TAE agarose gel, stained with ethidium bromide and observed under UV light. The PCR product was treated with ExoSAP-IT PCR clean up reagent (Thermo Fisher Scientific, Massachusetts, USA) and sequenced by the 3130xl Genetic Analyzer (Thermo Fisher Scientific) with the BigDye Terminator v3.1 Cycle Sequencing kit (Thermo Fisher Scientific). The nucleotide sequences were assembled and edited individually using DNASIS Pro (Hitachi Software Engineering, Tokyo, Japan).

The numbers of base differences per site among sequences (p-distance) were calculated and constructed Unweighted Pair Group Method using arithmetic Average (UPGMA) tree using the p-distance by Molecular Evolutionary Genetics Analysis software (MEGA X) [73].

Median-joining networks among firefly haplotypes were constructed and post-processed under maximum parsimony in Network Version 4.6.1.1 (available at http://fluxus-engineering.com/sharenet.htm) to describe phylogeographic and genetic relationships between haplotypes.

5.2.4 Flashing behavior analysis

The live adult fireflies from each population were brought to the laboratory (26°C) for recording flash patterns within 1–2 days after collection to decrease the error from weakness and death. They were paired 1: 1 for mating in a mating arena that was prepared from a 7.1 × 11.0 × 6.5 cm of transparent plastic box with small moist cotton. They were allowed to have an adaptation period for 15–30 min before

Figure 3.
Habitat characteristics of the firefly collection sites, a) SPK, b) PTE, c) PTE2, d) NPT, e) NPT2 and f) SPB.

starting the experiment. The experiment was carried out under dark conditions (0 lux) for 30 min to 2 hr. after sunset.

The flashing communication was recorded using a Sony Handycam™ digital camera recorder (HDR-SR11E) at nightshot mode. All experimental mating boxes were separated from one another by placing black partitions between each arena to prevent flash interference from other mating pairs. Ten to 15 mating pairs from each population were randomly selected for video recording. Two flash types, courtship and warning flash types (**Figure 4**), which appeared at different periods of mating sequences, were recorded. The "courtship flashes" produced during courtship in responding to females, perhaps displayed during dorsal mounting. On the other hand, the brighter flashes displayed mostly during copulation called were defines as "warning flashes." At least 15 sec intervals or 30–50 flashes were recorded from each male. In case of small populations that had low numbers of females, the males were allowed to mate with virgin captive females to stimulate courtship behavior.

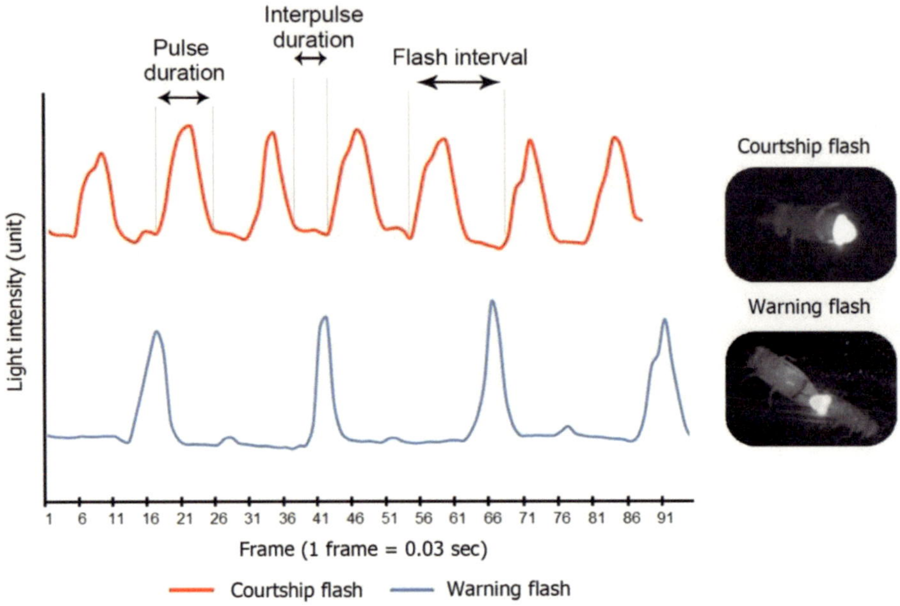

Figure 4.
Flashing behavior of male fireflies, courtship flash type (upper) and warning flash type (lower).

The video files were converted to audio video interleave or. AVI format files to analyze the flash parameters using time-lapse image analysis (TiLIA), a free software package for signal and flight pattern analyses of fireflies (available at Google Drive: https://drive.google.com/open?id=0B2o7FRVs2VohMmx2QzBVX3ZDeDA) [74] following the technique used by Thancharoen and Masoh [75]. The flash analysis was classified into three parameters, pulse duration, interpulse duration and flash interval, following previous study [76].

5.2.5 Statistical analysis

At least 30 flashes of courtship and warning flashes from each male were statistically analyzed. The pulse duration, interpulse duration, and flash interval among study sites were compared using One-way ANOVA and Tukey's multiple comparison tests. A value of $p < 0.05$ was considered statistically significant. The relationship between pulse and interpulse durations was tested using Pearson's correlation. All statistical analysis was performed using SPSS program version 24.

5.3 Results

5.3.1 Flashing behavior analysis

During mating behavior of *S. aquatilis*, the pulse durations of both courtship and warning flash types were quite similar, whereas the interpulse duration of warning flashes were twice longer than courtship flashes (**Table 2**). The correlation analysis of interpulse duration and pulse duration in each population showed that both flash parameters were negatively correlated (r in the range of −0.767 to −0.329, $P < 0.05$, n = 13). In case of short pulse duration, the interpulse duration was observed to be prolonged, stabilizing the flash interval.

The comparison of courtship flash parameters of all seven populations from five provinces showed that the fireflies from Suphan Buri province displayed different

Flash parameter	Duration in frame unit (mean ± SE)	
	Courtship flash (n = 60)	Warning flash (n = 28)
Pulse duration	5.54 ± 0.11	6.03 ± 0.17
Interpulse duration	6.78 ± 0.10	18.91 ± 0.34
Flash interval	12.32 ± 0.15	24.95 ± 0.38
Flash frequency	8.18 ± 0.09	4.03 ± 0.58

Table 2.
Flash parameters of courtship and warning flash types of S. aquatilis *(from overall populations).*

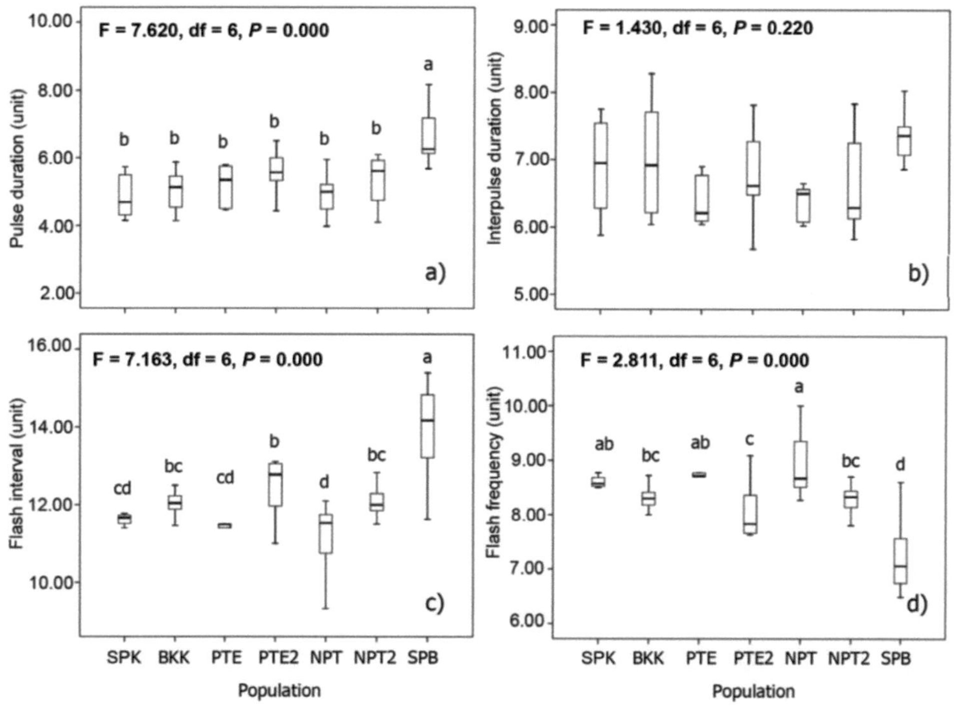

Figure 5.
The comparison of courtship flash parameters among seven populations of S. aquatilis; different letters indicate significant differences among different populations. Samut Prakarn (SPK), Bangkok (BKK), Pathum Thani (PTE), Nakhon Pathom (NPT), and Suphan Buri (SPB).

courtship flashes from the other sites located in the Bangkok Metropolitan Region (Samut Prakarn, Pathum Thani, Nakhon Pathom and Bangkok) (One-way ANOVA, $P < 0.05$; **Figure 5**). Results indicated that the Suphan Buri population had significantly longer pulse duration and flash interval resulting in slow flashing.

The flash parameters of the warning flash type could not be analyzed in all populations because not all experimental mating pairs displayed warning flashes. Therefore, only three populations from Pathum Thani, Nakhon Pathom and Suphan Buri province were analyzed. Perhaps because the mating happened under controlled environments without interference from mate competition and predation. Again, the Suphan Buri population flashed significantly differed when compared with other populations (**Figure 6**). It had a significantly long interpulse duration that resulted in having a long flash interval and a low flash frequency.

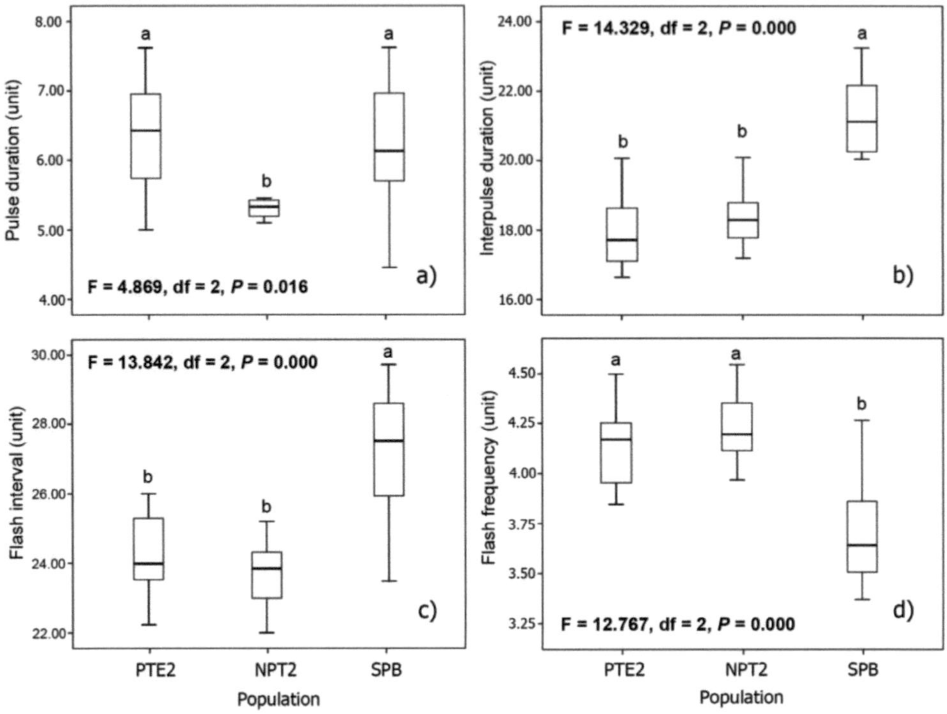

Figure 6.
The comparison of warning flash parameters among three populations of S. aquatilis; different letters indicate significant differences among different populations. Samut Prakarn (SPK), Bangkok (BKK), Pathum Thani (PTE), Nakhon Pathom (NPT), and Suphan Buri (SPB).

5.3.2 Genetic diversity of S. aquatilis

The genetic diversity of COII gene in *S. aquatilis* populations were examined from 132 individuals from seven locations in five provinces in the central part of Thailand. The sequences were registered in GenBank accession nos. MW800771 to MW800823 and MW814512 to MW814587. The p-distances among individuals ranged from 0 to 0.0122. The UPGMA tree revealed that regional cohesion of sequence types was not observed due to short p-distances (data not shown). The median-joining haplotype network was needed to confirm the low genetic diversity. The network revealed 37 haplotypes but not any phylogeographic sub-structuring of the firefly populations (**Figure 7**). Thus, no genetic differentiation was shown among the *S. aquatilis* populations examined.

5.4 Discussion

The study revealed flash signal variation among populations of *S. aquatilis* in the central part of Thailand. However, a distant population in Suphan Buri province apparently displayed longer pulse duration in the courtship flashes and longer interpulse in the warning flashes. As sexual communication, the pulse duration of the courtship signals is generally quite similar, preserving constant species-specific flash patterns. Most researchers studied "interflash interval" to define flash type from frequency, for instance, slow-flash, fast-flash, intermediate-flash and quick flash types [65–68]. However, the negative correlation between interpulse duration and pulse duration might help to balance the flash interval and flash frequency.

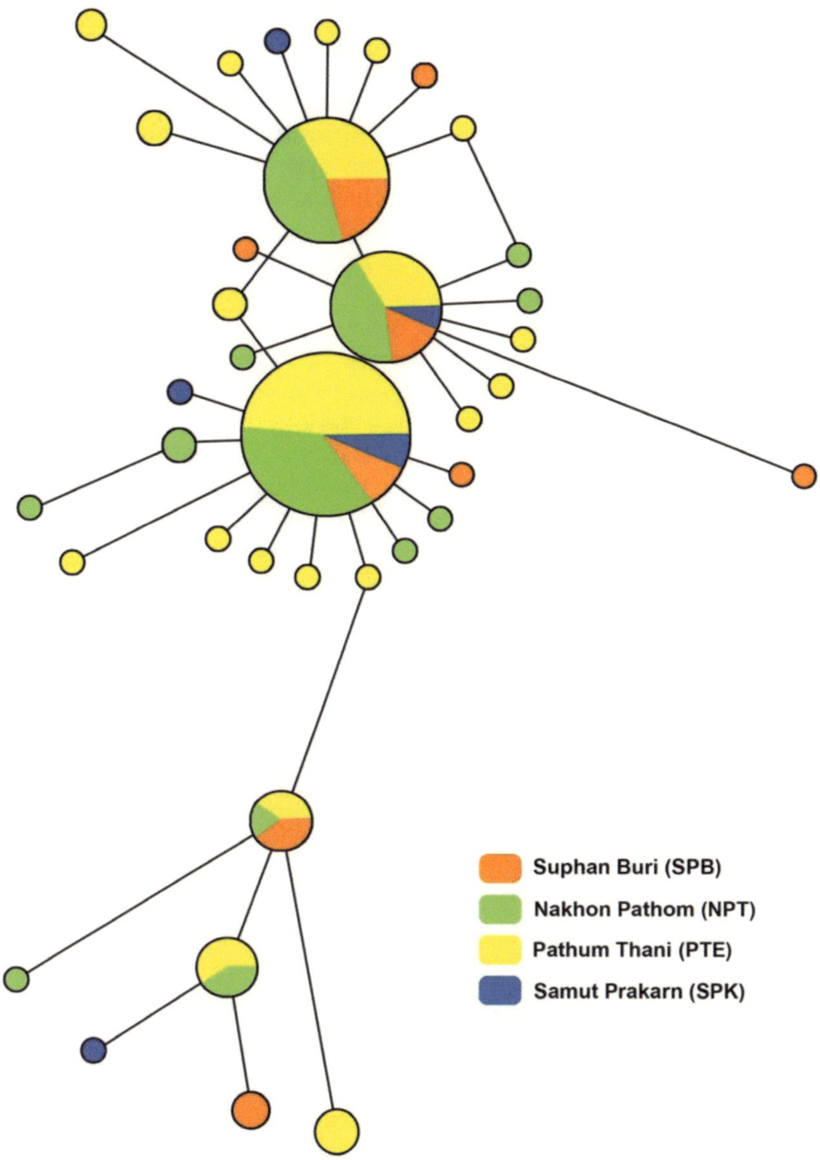

Figure 7.
Median-joining haplotype network generated from COII data from S. aquatilis collected from four locations in Central Thailand, different colors represent different collecting locations, sizes of nodes and pie segments are proportional to haplotype frequency, and length of branches is proportional to number of mutational changes between haplotypes.

Our finding was that there is intraspecific variation in flash communication of *S. aquatilis*. The fireflies in the Bangkok Metropolitan Region were fast-flash populations whereas the Suphan Buri population was slow-flashing although they did not show genetic differences among populations. This result is similar to the case of *L. lateralis* that *L. lateralis* populations distributed throughout the Korean Peninsula, Northeast China, Sakhalin, and Japan, the two flashing behavioral types could not be separated phylogenetically [69]. However, among populations with different flash types of *L. cruciata* in Japan, the genetic variation associated with flashing behavior was investigated [65, 67, 68]. The geographical differences caused by a great rupture zone of Japanese Islands might have had a strong effect on this species. Similarly, as the most geographically distant location of our studied populations, the Suphan Buri population (109 kilometers from Bangkok), is probably isolated from the others. Although there are no geographical barriers influencing allopatric populations like in the Japanese case, habitat fragmentation including light pollution barriers probably significantly affect the firefly populations. *S. aquatilis* fireflies normally inhabit in or near freshwater areas, the active males can fly fast and travel a long distance, the inactive females remain near a water area. The reduced female mobility behavior might limit the dispersal ability of the species and result in population isolation. In addition, artificial night lighting could also interfere with flashes of *S. aquatilis* resulting in adaptive behavior to adjust their flashes.

The fireflies inhabiting the area of the Bangkok Metropolitan Region might face a habitat flooded with artificial light that causes reduced ability to communicate with their mates. Selection pressure favors adaptations of their flash pattern to minimize light competition or to increase the clarity of flash signals to improve their mating success. It might be possible that the environmental selection pressure happened in the fireflies. The plasticity of the flashing behavior depending on situation and environmental conditions were examined in many firefly species [75, 77, 78]. The fireflies in light polluted areas will modify their flash patterns to be faster to mitigate steady light from artificial night lighting. Similar adaptations occur in acoustically communicating animals, where ambient noise, especially anthropogenic low-frequency noise, affected acoustic communication in blackbirds [79], tree frogs [80], tree swallows [81], fish [82] and tree crickets [83]. The birds sing louder with higher frequencies to mitigate low frequency traffic noise, while the males of the tree crickets shortened their calls (echemes) and paused singing with a higher probability with increasing noise level without modification of song frequency or interecheme interval. Unfortunately, no work has been done on their genetic differences between the normal and noise polluted populations.

5.5 Recommendations

Generally, genetic differentiation among populations would happen in a heterogenous or mosaic environment by reduction of population size, genetic drift, gene flow and natural selection and accumulated by geographic isolation. Although there is no geographical isolation in the central region of Thailand, in case of *S. aquatilis*, gene flow is limited by the dispersal ability of adult females and aquatic larvae that are restricted to the aquatic ecosystems. In addition, the light pollution is likely an important barrier limiting the adult dispersal whereas habitat fragmentation reduces population sizes, reduces habitat size of firefly larvae and increases isolation of small subpopulations. The wild populations of the fireflies are at risk of extinction due to the effect of inbreeding depression.

The recommendation for *S. aquatilis* translocation is to consider: (i) no genetic differentiation between the local and the released populations, (ii) no divergence in flash signals to prevent pre-mating isolation between recipient and donor populations, (iii) the distance between populations might promote variation among populations; thus, closer populations are properly used for translocation, (iv) the sources of translocated populations come from a large population or several subpopulations to acquire proper numbers of source populations and decrease the effect of inbreeding depression. In addition, other factors, for example, habitat quality, source of translocated fireflies (from wild or captivity), released stage, frequency of releasing, released area and other environmental conditions during releasing, can relate to the success of program. This information is probably species specific; therefore, the biological and ecological characteristics of the focus species are needed for translocation application. Significantly, the long-term monitoring of establish populations also is necessary.

In the case study, although the *S. aquatilis* populations in the central part of Thailand have no genetic divergence among populations, the variation of flash signals was found in a location of Suphan Buri province. The translocation of the species could happen if the donor and recipient populations come from Bangkok Metropolitan Region where the fireflies displayed similar flash signals and no genetic divergence among populations.

Acknowledgements

I would like to thank Dr. Praneet Damrongphol, Mahidol University for providing advice on the experimental designs and constructive comments on this manuscript, Dr. Hirobumi Suzuki, Olympus Corporation for technical assistance, Dr.Yuvarin Boontop, Department of Agriculture, Bangkok for molecular data analysis and Dr. Parichart Laksanawimol, Chandrakasem Rajabhat University for field assistant. I especially thank Dr. Steven Wing for manuscript correction.

This study was financially supported by the Thailand Research Fund (TRF) (MRG5580177) and Kasetsart University Research and Development Institute (KURDI).

Conflict of interest

The author declares no conflict of interest.

Author details

Anchana Thancharoen
Department of Entomology, Faculty of Agriculture, Kasetsart University, Bangkok, Thailand

Address all correspondence to: agrant@ku.ac.th

References

[1] Buck J, Buck E. Synchronous fireflies. Scientific American. 1976;234(5):74-85.

[2] Thancharoen A. Well managed firefly tourism: A good tool for firefly conservation in Thailand. Lampyrid. 2012;2:142-148.

[3] Nada B, Kirton LG, Khoo V. Conservation efforts for the synchronous fireflies of the Selangor River in Malaysia. In: Proceedings of International Firefly Symposium on Diversity and Conservation of Fireflies; Chiang Mai, Thailand; 2008

[4] Nallakumar K. The synchronously flashing aggregative fireflies of Peninsular Malaysia. Biodiversity. 2003;4(2):11-16. DOI:10.1080/14888386.2003.9712684

[5] Napompeth, B. Reminiscience of firefly study in Thailand. In: Proceedings of International Firefly Symposium on Diversity and Conservation of Fireflies; Chiang Mai, Thailand; 2008

[6] Lewis SM, Wong CH, Owens A, Fallon C, Jepsen S, Thancharoen A, Wu C, De Cock R, Novák M, López-Palafox T, Khoo V. A global perspective on firefly extinction threats. BioScience. 2020;70(2):157-167. DOI: 10.1093/biosci/biz157

[7] Owens AC, Lewis SM. The impact of artificial light at night on nocturnal insects: A review and synthesis. Ecology and evolution. 2018;8(22):11337-11358. DOI: 10.1002/ece3.4557

[8] Harvey JA, Heinen R, Armbrecht I, Basset Y, Baxter-Gilbert JH, Bezemer TM, Böhm M, Bommarco R, Borges PA, Cardoso P, Clausnitzer V. International scientists formulate a roadmap for insect conservation and recovery. Nature Ecology & Evolution. 2020;4(2):174-176. DOI: 10.1038/s41559-019-1079-8

[9] Cheng S, Faidi MA, Tan SA, Vijayanathan J, Malek MA, Bahashim B, Isa MN. Fireflies in Southeast Asia: knowledge gaps, entomotourism and conservation. Biodiversity and Conservation. 2021;5:1-20. DOI: 10.1007/s10531-021-02129-3

[10] Lewis SM, Thancharoen A, Wong CH, et al. Firefly tourism: Advancing a global phenomenon toward a brighter future. Conservation Science and Practice. 2021;e391. https://doi.org/10.1111/csp2.391

[11] Fu X, Nobuyoshi O, Zhang Y, Lei C. A rearing apparatus and diet for the aquatic firefly *Luciola leii* (Coleoptera: Lampyridae). Canadian entomologist. 2006;138(3):399. DOI: 10.4039/n05-029

[12] Ho JZ, Chiang PH, Wu CH, Yang PS. Life cycle of the aquatic firefly *Luciola ficta* (Coleoptera: Lampyridae). Journal of Asia-Pacific Entomology. 2010;13(3):189-196. DOI: 10.1016/j.aspen.2010.03.007

[13] Jeng ML, Lai J, Yang PS. Lampyridae: a synopsis of aquatic fireflies with description of a new species (Coleoptera). Water beetles of China. 2003;3:539-562.

[14] Kang SH, Jeon MK, Kwon SJ, Na SJ, Kim KH, Jeong JC. Artificial Habitat Creation of *Luciola lateralis* (Coleoptera: Lamphyridae) and Research of breeding technique for festival at Hwadamsup, Korea. Journal of forest and environmental science. 2018;34(4):275-283.

[15] Thancharoen A. The apparatus for egg oviposition and hatching of aquatic fireflies. Thai Patent No. 15274. 2019a.

[16] Thancharoen A. The apparatus for rearing aquatic firefly larvae. Thai Patent No. 15275. 2019b.

[17] Fu X, Wang Y, Lei C, Nobuyoshi O. The swimming behavior of the aquatic larvae of the firefly *Luciola substriata* (Coleoptera: Lampyridae). The Coleopterists Bulletin. 2005;59(4):501-505. DOI: 10.1649/830.1

[18] Chen JC, Chang WL. Study of public participation on ecological restoration for firefly habitat –Muzha Park's Cui Lake in Taipei In: Proceedings of International Firefly Symposium; 23-28 April 2017; Taipei, Taiwan. p. 36

[19] Wu BW, Chang WL. Study of Ecological Compensation of Firefly Ditch in Taipei Muzha In: Proceedings of International Firefly Symposium; 23-28 April 2017; Taipei, Taiwan. p. 37

[20] Tsai Mc, Chang WL. *Aquatica ficta* (Olivier) Habitat conservation in Daan Forest Park of Taipei In: Proceedings of International Firefly Symposium; 23-28 April 2017; Taipei, Taiwan; p. 38

[21] Wu C and Yang PS. How to design and maintain an aquatic firefly eco-pond in conservation biology, ecology and engineering principles In: Proceedings of International Firefly Symposium; 23-28 April 2017; Taipei, Taiwan; p. 40

[22] Huang L. The key role of the fireflies restoration in Taipei City – The Government Department In: Proceedings of International Firefly Symposium; 23-28 April 2017; Taipei, Taiwan; p. 41

[23] Kang SH, Jeon MK, Kwon SJ, Na SJ, Kim KH, Jeong JC. Artificial Habitat Creation of *Luciola lateralis* (Coleoptera: Lamphyridae) and Research of Breeding Technique for Festival at Hwadamsup, Korea. Journal of forest and environmental science. 2018;34(4):275-283.

[24] Waley P. Symbol, space and ecosystem in the waterways of Japan. Animal Spaces, Beastly Places: New geographies of human-animal relations. 2000;10:159.

[25] Iguchi Y. The ecological impact of an introduced population on a native population in the firefly *Luciola cruciata* (Coleoptera: Lampyridae). Biodiversity and conservation. 2009;18(8):2119-2126. DOI: 10.1007/s10531-009-9576-8

[26] Kato DI, Suzuki H, Tsuruta A, Maeda J, Hayashi Y, Arima K, Ito Y, Nagano Y. Evaluation of the population structure and phylogeography of the Japanese Genji firefly, *Luciola cruciata*, at the nuclear DNA level using RAD-Seq analysis. Scientific reports. 2020;10(1):1-2. DOI: 10.1038/s41598-020-58324-9

[27] IUCN/SSC. Guidelines for Reintroductions and Other Conservation Translocations. Version 1.0, IUCN Species Survival Commission. Gland, Switzerland: IUCN Species Survival Commission, viiii; 2013. 57 p. Available from: https://www.iucn.org/content/guidelines-reintroductions-and-other-conservation-translocations [Accessed: 2021-03-07]

[28] Corlett RT. Restoration, reintroduction, and rewilding in a changing world. Trends in ecology & evolution. 2016;31(6):453-462. DOI: 10.1016/j.tree.2016.02.017

[29] Olech W, Perzanowski K. A genetic background for reintroduction program of the European bison (*Bison bonasus*) in the Carpathians. Biological Conservation. 2002;108(2):221-228. DOI: 10.1016/s0006-3207(02)00108-8

[30] Drauch AM, Rhodes Jr OE. Genetic evaluation of the lake sturgeon reintroduction program in the Mississippi and Missouri Rivers. North American Journal of Fisheries Management. 2007;27(2):434-442. DOI: 10.1577/m06-024.1

[31] Nielsen RK, Pertoldi C, Loeschcke V. Genetic evaluation of the captive breeding program of the Persian wild ass. Journal of zoology. 2007;272(4):349-357. DOI: 10.1111/j.1469-7998.2007.00294.x

[32] Stockwell MP, Clulow S, Clulow J, Mahony M. The impact of the amphibian chytrid fungus *Batrachochytrium dendrobatidis* on a green and golden bell frog *Litoria aurea* reintroduction program at the Hunter Wetlands Centre Australia in the Hunter Region of NSW. Australian Zoologist. 2008;34(3):379-386. DOI: 10.7882/az.2008.015

[33] Hedrick PW, Fredrickson RJ. Captive breeding and the reintroduction of Mexican and red wolves. Molecular Ecology. 2008;17(1):344-350. DOI: 10.1111/j.1365-294x.2007.03400.x

[34] Hannon ER, Hafernik JE. Reintroduction of the rare damselfly *Ischnura gemina* (Odonata: Coenagrionidae) into an urban California park. Journal of insect conservation. 2007;11(2):141-149. DOI: 10.1007/s10841-006-9027-8

[35] Hochkirch A, Witzenberger KA, Teerling A, Niemeyer F. Translocation of an endangered insect species, the field cricket (*Gryllus campestris* Linnaeus, 1758) in northern Germany. In: Biodiversity and Conservation in Europe 2006 (pp. 355-365). Springer, Dordrecht. DOI: 10.1007/978-1-4020-6865-2_25

[36] Griffith B, Scott JM, Carpenter JW, Reed C. Translocation as a species conservation tool: status and strategy. Science. 1989;245(4917):477-480. DOI: 10.1126/science.245.4917.477

[37] Miskelly CM, Powlesland RG. Conservation translocations of New Zealand birds, 1863-2012. Notornis. 2013;60(1):3-28. DOI: 10.1525/9780520930636-009

[38] Katherine R, Jonathan BD. Captive Breeding and Reintroduction. In: Levin S.A. (ed.) Encyclopedia of Biodiversity, 2nd edition, 2013 Volume 1, pp. 662-667. Waltham, MA: Academic Press. DOI: 10.1016/b0-12-226865-2/00041-9

[39] Robert A. Captive breeding genetics and reintroduction success. Biological Conservation. 2009;142(12):2915-2922. DOI: 10.1016/j.biocon.2009.07.016

[40] Christie MR, Marine ML, French RA, Blouin MS. Genetic adaptation to captivity can occur in a single generation. Proceedings of the National Academy of Sciences. 2012;109(1):238-242. DOI: 10.1073/pnas.1111073109

[41] Williams SE, Hoffman EA. Minimizing genetic adaptation in captive breeding programs: a review. Biological conservation. 2009;142(11):2388-2400. DOI: 10.1016/j.biocon.2009.05.034

[42] Theodorou K, Couvet D. The efficiency of close inbreeding to reduce genetic adaptation to captivity. Heredity. 2015;114(1):38-47. DOI: 10.1038/hdy.2014.63

[43] Kasso M, Balakrishnan M. Ex situ conservation of biodiversity with particular emphasis to Ethiopia. International Scholarly Research Notices. 2013;2013. DOI: 10.1155/2013/985037

[44] Weeks AR, Sgro CM, Young AG, Frankham R, Mitchell NJ, Miller KA, Byrne M, Coates DJ, Eldridge MD, Sunnucks P, Breed MF. Assessing the benefits and risks of translocations in changing environments: a genetic perspective. Evolutionary Applications. 2011;4(6):709-725. DOI: 10.1111/j.1752-4571.2011.00192.x

[45] Tarszisz E, Dickman CR, Munn AJ. Physiology in conservation

translocations. Conservation Physiology. 2014;2(1). DOI: 10.1093/conphys/cou054

[46] Fraser DJ. How well can captive breeding programs conserve biodiversity? A review of salmonids. Evolutionary Applications. 2008;1(4):535-586. DOI: 10.1111/j.1752-4571.2008.00036.x

[47] Frankham R. Genetic adaptation to captivity in species conservation programs. Molecular ecology. 2008;17(1):325-333. DOI: 10.1111/j.1365-294x.2007.03399.x

[48] Polak T, Saltz D. Reintroduction as an ecosystem restoration technique. Conservation Biology. 2011;25(3):424-427. DOI: 10.1111/j.1523-1739.2011.01669.x

[49] Hodder, Kathy H., and James M. Bullock. "Translocations of Native Species in the UK: Implications for Biodiversity." Journal of Applied Ecology. 1997; 34(3): 547-565. DOI:10.2307/2404906.

[50] Pearce-Kelly P, Jones R, Clarke D, Walker C, Atkin P, Cunningham AA. The captive rearing of threatened Orthoptera: a comparison of the conservation potential and practical considerations of two species' breeding programmes at the Zoological Society of London. Journal of Insect Conservation. 1998;2(3):201-210.

[51] Bellis J, Bourke D, Williams C, Dalrymple S. Identifying factors associated with the success and failure of terrestrial insect translocations. Biological Conservation. 2019;236:29-36. DOI: 10.1016/j.biocon.2019.05.008

[52] Swan KD, Lloyd NA, Moehrenschlager A. Projecting further increases in conservation translocations: a Canadian case study. Biological Conservation. 2018;228:175-182. DOI: 10.1016/j.biocon.2018.10.026

[53] Hodder KH, Bullock JM. Translocations of native species in the UK: implications for biodiversity. Journal of Applied Ecology. 1997:547-565. DOI: 10.2307/2404906.

[54] Witzenberger KA, Hochkirch A. Genetic consequences of animal translocations: A case study using the field cricket, *Gryllus campestris* L. Biological Conservation. 2008;141(12):3059-3068. DOI: 10.1016/j.biocon.2008.09.017

[55] Knisley CB, Hill JM, Scherer AM. Translocation of threatened tiger beetle *Cicindela dorsalis dorsalis* (Coleoptera: Cicindelidae) to Sandy hook, New Jersey. Annals of the Entomological Society of America. 2005;98(4):552-557. DOI: 10.1603/0013-8746(2005)098[0552:TOTTBC]2.0.CO;2

[56] Hannon ER, Hafernik JE. Reintroduction of the rare damselfly *Ischnura gemina* (Odonata: Coenagrionidae) into an urban California park. Journal of insect conservation. 2007;11(2):141-149. DOI: 10.1007/s10841-006-9027-8

[57] Longcore T, Bonebrake T. Captive propagation and release plan for Quino Checkerspot butterfly (*Euphydryas editha quino*). 2012. Available from: https://www.researchgate.net/profile/Travis-Longcore/publication/334602577_Captive_Propagation_and_Release_Plan_for_Quino_Checkerspot_Butterfly_Euphydryas_editha_quino/links/5d3512384585153e59167482/Captive-Propagation-and-Release-Plan-for-Quino-Checkerspot-Butterfly-Euphydryas-editha-quino.pdf [Accessed: 2021-03-07]

[58] Haugan EB. 'Homeplace of the Heart': Fireflies, Tourism and Town-Building in Rural Japan (Master's thesis). Oslo: University in Oslo, Norway; 2019. Available from: https://www.duo.uio.no/handle/10852/70274 [Accessed: 2021-03-07]

[59] Schultz CB, Russell C, Wynn L. Restoration, reintroduction, and captive propagation for at-risk butterflies: a review of British and American conservation efforts. Israel Journal of Ecology and Evolution. 2008;54(1):41-61. DOI: 10.1560/ijee.54.1.41

[60] Boppré M, Vane-Wright RI. The butterfly house industry: conservation risks and education opportunities. Conservation and Society. 2012;10(3):285-303. DOI: 10.4103/0972-4923.101831

[61] Lewis OT, Thomas CD. Adaptations to captivity in the butterfly *Pieris brassicae* (L.) and the implications for ex situ conservation. Journal of Insect Conservation. 2001;5(1):55-63.

[62] Usener JL, Cognato AI. Patterns of mitochondrial diversity among desert firefly populations (Lampyridae: *Microphotus octarthrus* Fall). The Coleopterists Bulletin. 2005;59(3):361-367. DOI: 10.1649/796.1

[63] Lee SC, Bae JS, Kim I, Suzuki H, Kim SR, Kim JG, Kim KY, Yang WJ, Lee SM, Sohn HD, Jin BR. Mitochondrial DNA sequence-based population genetic structure of the firefly, *Pyrocoelia rufa* (Coleoptera: Lampyridae). Biochemical genetics. 2003;41(11):427-452. DOI: 10.1023/b:bigi.0000007777.87407.1b

[64] Suzuki H, Sato Y, Fujiyama S, Ohba N. Genetic differentiation between ecological two types of the Japanese firefly, *Hotaria parvula*: An electrophoretic analysis of allozymes. Zool. Sci. 1993;10:697-703.

[65] Suzuki H, Sato Y, Fujiyama S, Ohba N. Biochemical systematics of Japanese fireflies of the subfamily Luciolinae and their flash communication systems. Biochemical genetics. 1996;34(5):191-200.

[66] Tamura M, Yokoyama J, Ohba N, Kawata M. Geographic differences in flash intervals and pre-mating isolation between populations of the Genji firefly, *Luciola cruciata*. Ecological entomology. 2005;30(2):241-245. DOI: 10.1111/j.0307-6946.2005.00683.x

[67] Suzuki H, Sato Y, Ohba N. Gene diversity and geographic differentiation in mitochondrial DNA of the Genji firefly, *Luciola cruciata* (Coleoptera: Lampyridae). Molecular Phylogenetics and Evolution. 2002 22(2):193-205. DOI: 10.1006/mpev.2001.1046

[68] Ohba SY, Numata K, Kawano K. Variation in flash speed of Japanese firefly, *Luciola cruciata* (Coleoptera: Lampyridae), identifies distinct southern "quick-flash" population on Goto Islands, Japan. Entomological Science. 2020;23(2):119-127. DOI: 10.1111/ens.12403

[69] Suzuki H, Sato Y, Ohba N, Bae JS, Jin BR, Sohn HD, Kim SE. Phylogeographic analysis of the firefly, *Luciola lateralis*, in Japan and Korea based on mitochondrial cytochrome oxidase II gene sequences (Coleoptera: Lampyridae). Biochemical genetics. 2004;42(9):287-300. DOI: 10.1023/b:bigi.0000039805.75118.8f

[70] Thancharoen A, Ballantyne LA, Branham MA, Jeng ML. Description of *Luciola aquatilis* sp. nov., a new aquatic firefly (Coleoptera: Lampyridae: Luciolinae) from Thailand. Zootaxa. 2007 Oct 10;1611(1):55-62. DOI: 10.11646/zootaxa.1611.1.4

[71] Thancharoen A. The biology and mating behavior of an aquatic firefly species, *Luciola aquatilis* sp. nov. Thancharoen (Coleoptera: Lampyridae). Bangkok: Mahidol University; 2007.

[72] Saccheri I, Kuussaari M, Kankare M, Vikman P, Fortelius W, Hanski I. Inbreeding and extinction in a butterfly metapopulation. Nature. 1998;392(6675):491-494. DOI: 10.1038/33136

[73] Kumar S, Stecher G, Li M, Knyaz C, Tamura K: MEGA X: Molecular Evolutionary Genetics Analysis across computing platforms. Molecular Biology and Evolution. 2018;35:1547-1549 DOI: 10.1093/molbev/msy096

[74] Konno J, Hatta-Ohashi Y, Akiyoshi R, Thancharoen A, Silalom S, Sakchoowong W, Yiu V, Ohba N, Suzuki H. TiLIA: A software package for image analysis of firefly flash patterns. Ecology and evolution. 2016;6(9):3026-3031. DOI: 10.1002/ece3.2078

[75] Thancharoen A, Masoh S. Effect of camera illumination on flashing behavior of *Pteroptyx malaccae* (Coleoptera: Lampyridae). In: Hirobumi Suzuki, editor. Bioluminescence-Analytical Applications and Basic Biology: IntechOpen; 2019, DOI: 10.5772/intechopen.85796.

[76] Lewis SM, Cratsley CK, Demary K. Mate recognition and choice in Photinus fireflies. In: Annales Zoologici Fennici 2004 Jan 1 (pp. 809-821). Finnish Zoological and Botanical Publishing Board.

[77] Carlson AD, Copeland J. Behavioral plasticity in the flash communication systems of fireflies: Although insect behavior has generally been considered stereotyped, recent research indicates that fireflies can alter their flash patterns according to the behavioral context. American Scientist. 1978;66(3):340-346.

[78] Owens ACS, Meyer-Rochow VB, Yang EC. Short- and mid-wavelength artificial light influences the flash signals fireflies (Coleoptera: Lampyridae). PLOS One. 2018;13(2):e0191576. DOI: 10.1371/journal.pone.0191576

[79] Nemeth E, Brumm H. Blackbirds sing higher-pitched songs in cities: adaptation to habitat acoustics or side-effect of urbanization?. Animal behaviour. 2009;78(3):637-641. DOI: 10.1016/j.anbehav.2009.06.016.

[80] Lengagne T. Traffic noise affects communication behaviour in a breeding anuran, *Hyla arborea*. Biological conservation. 2008;141(8):2023-2031. DOI: 10.1016/j.biocon.2008.05.017

[81] McIntyre E, Leonard ML, Horn AG. Ambient noise and parental communication of predation risk in tree swallows, *Tachycineta bicolor*, Animal Behaviour. 2014;87:85-89. DOI: 10.1016/j.anbehav.2013.10.013

[82] Radford AN, Kerridge E, Simpson SD. Acoustic communication in a noisy world: can fish compete with anthropogenic noise?. Behavioral Ecology. 2014;25(5):1022-1030. DOI: 10.1093/beheco/aru029

[83] Orci KM, Petróczki K, Barta Z. Instantaneous song modification in response to fluctuating traffic noise in the tree cricket *Oecanthus pellucens*. Animal Behaviour. 2016;112:187-194. DOI: 10.1016/j.anbehav.2015.12.008

Chapter 6

Semi-Intrinsic Luminescence in Marine Organisms

Jeremy Mirza and Yuichi Oba

Abstract

Light emission is widespread in the oceans, with over three quarters of all observed marine species exhibiting bioluminescence. Several organisms such as the copepod *Metridia pacifica* and the ostracod *Vargula hilgendorfii* have been proven to synthesise their luciferin and luciferase to facilitate light emission. However, many luminescent species lack the capability to do this and instead it is possible that they acquire some of the components for their luminescence through predation or filter feeding on organisms that produce luciferins or precursors to these molecules. This has resulted in many organisms using certain luciferins, such as coelenterazine, as their substrate without possessing a clear mechanism to synthesise these. This chapter will review several examples of these semi-intrinsic luminescent systems and how the substrates and enzymes can be obtained for these reactions. Moreover, it will look at why particular luciferins, such as coelenterazine, are more widespread and utilised in this manner compared to other substrates.

Keywords: Bioluminescence, Semi-Intrinsic, Luciferin, Coelenterazine, Imidazopyrazinone

1. Introduction

Bioluminescence is a chemical process numerous organisms utilise to produce light. This reaction has been studied in a wide range of taxa, in terms of its chemistry, evolutionary history and purpose in ecology [1]. This ability to emit light via a chemical reaction can be found in a diverse range of phyla, ranging from simple unicellular bacteria and protists to more complex organisms such as cephalopods and elasmobranchs [1]. Generally, this is a chemical reaction that involves the oxidation of a luciferin compound in the presence of a luciferase enzyme. This produces an unstable intermediate (usually a cyclic peroxide) that breaks down to produce a compound generically called oxyluciferin and gives off a large amount of energy as light [2, 3].

This phenomenon has evolved independently at least 94 and potentially over 100 times [4] across both marine and terrestrial genera, and around 80% of bioluminescent genera occur in the oceans [5, 6]. In marine ecosystems, it is estimated that up to 95% of organisms that dwell below 200 m depth are able to emit light [7–9]. Given the widespread utilisation of this phenomenon, there are a diverse array of luminescent systems that exist with several different substrates and a wide variety of associated enzymes.

Unlike the enzymatic component of the reaction where individual species are capable of expressing unique enzymes, luciferins are more conserved, and the same structures can be found across multiple distinct phyla. As of now at least 10 natural luciferins have been identified in terms of their chemical structure [4, 10]. Of those, the four main marine groups of luciferins are bacterial luciferin, tetrapyrrole used by dinoflagellates and krill, cypridinid luciferin used by several species of fish and ostracods and coelenterazine which is used by luminescent organisms in at least 9 different phyla [11].

Despite being a critical component for light emission, many marine organisms do not produce their own luciferins, and obtain these small organic compounds from their diet by grazing or predating on other luminescent organisms [1]. These species exhibit semi-intrinsic luminescence, as they still express their own luciferase enzymes, however they can obtain the substrates and potentially precursors to luciferin needed for luminescence through their diets [12]. Some have even shown the capacity to obtain the enzymatic component of the luminescent reaction through their diet as well [13]. With regards to this phenomenon the most notable examples of semi-intrinsic luminescence involve coelenterazine and cypridinid luciferin [14].

This chapter will review the prevalence of known semi-intrinsic luminescent systems and how these organisms have attained light emission. Moreover, it will look at why these reactions and predator–prey relationships have evolved over time and discuss why certain substrates are more commonly observed in semi-intrinsic luminescence.

2. Sources of luminescence in semi-intrinsic systems

Identifying the presence of luminescence in an organism is well established and involves identifying the luciferin and luciferase involved in the reaction and separating them. The basic technique for luciferin and luciferase separation, developed by Dubois [15, 16] is termed "hot-cold extract". In this method, two water extracts of luminogenic tissue are prepared [3, 16]. The use of cold extract allows to preserve the activity of the enzyme (luciferase), while the heated fraction destroys the proteins and yields the luciferin, and when both extracts are mixed together an *in vitro* luminescence is produced [3, 16]. Each extract can be purified to allow for the identification of the amino acid sequence corresponding to the luciferase and the chemical structure of the luciferin [3, 17].

However, this in of itself does not establish how the luminescent organism obtained these components. A possible method to identify this is by constructing the transcriptome of an organism to prove the luciferase enzyme was expressed and not obtained through diet [3, 18]. However, this is a lot more difficult when it comes to identifying whether an organism can synthesise its own luciferin, as very few biosynthetic pathways have been established.

Despite this, it has been shown by controlling the diet of a number of higher taxa that their luminescence is dependent on the consumption of particular organisms [12, 19]. Subsequently, it has been possible to identify several organisms at lower trophic levels that can produce their own luciferin, including the ostracod *Vargula hilgendorfii* [20] and the copepod *Metridia pacifica* [14], both shown in **Figure 1**.

2.1 Cypridinid luciferin

Cypridinid luciferin was the first marine luminescent substrate to be identified in terms of its chemical structure. This compound was first isolated and crystallised by Shimomura and colleagues [21, 22], and the structure was determined by Kishi et al. [23], allowing for the detailed study of the biochemistry of this reaction [1]. The ostracod *V. hilgendorfii* was shown to secrete a luminescent mucus when disturbed, emitting a bright blue light at a peak wavelength of 453–455 nm [24]. The

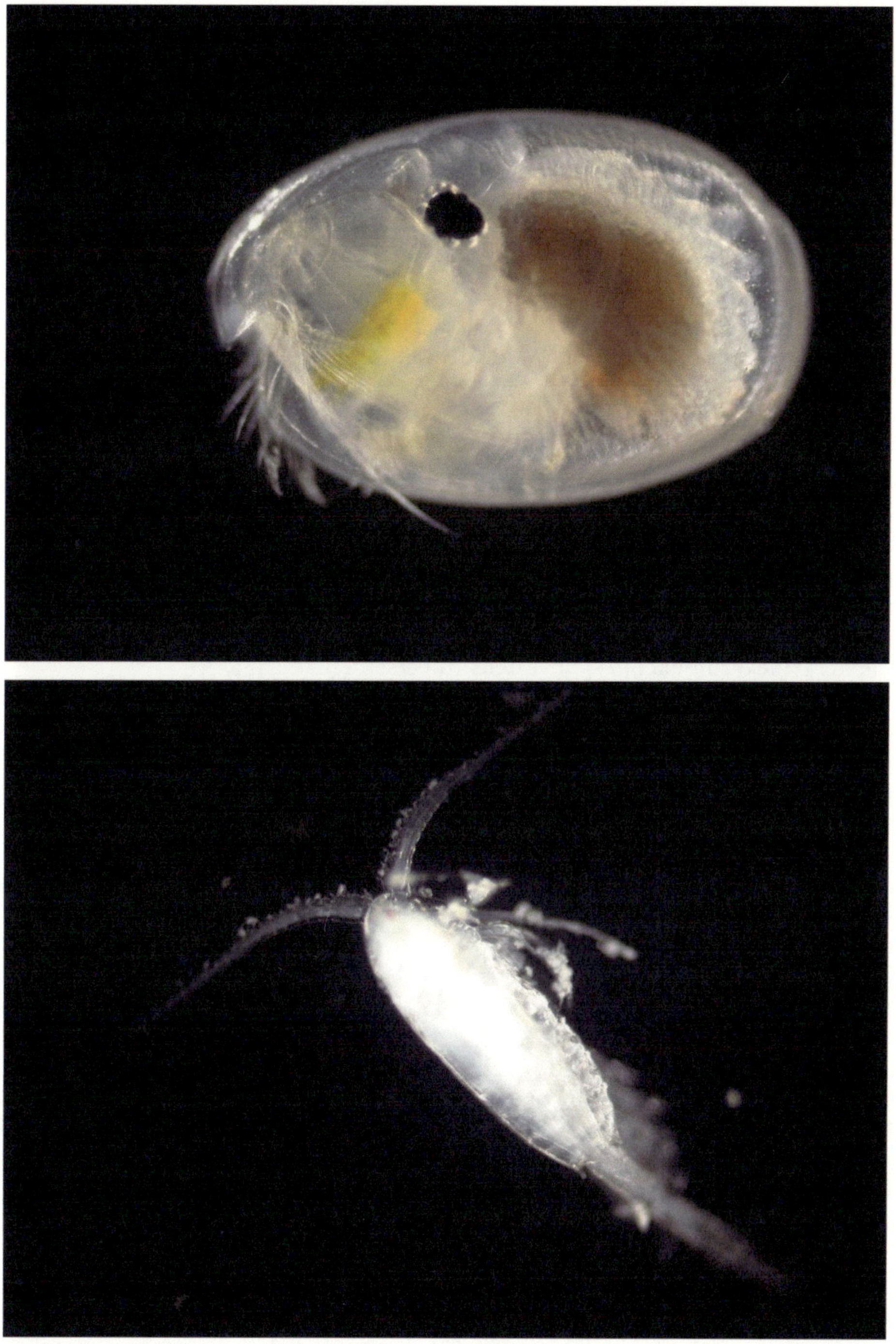

Figure 1.
Photographs of luminescent organisms known to synthesise their luciferins. The ostracod Vargula hilgendorfii *(upper) synthesises cypridinid luciferin and the copepod* Metridia pacifica *(lower) synthesises coelenterazine. Photos taken by ken-ichi Onodera, and Yuichi Oba.*

luminescent cloud of mucus is emitted from specialised glands from two types of cell, one producing the luciferin and the other the luciferase [25].

Kato and colleagues [26, 27] showed that ostracod luciferin is synthesised from tryptophan, isoleucine, and arginine, via a currently unknown pathway. This was observed by labelling the amino acid L-tryptophan with deuterium before feeding the ostracod *V. hilgendorfii* with this to confirm incorporation into the cypridinid luciferin [20]. *V. hilgendorfii* was shown to be the first example of a species that could use free amino acids to synthesise its imidazopyrazinone-type substrate, cypridinid luciferin. While this is used by several bioluminescent species, it makes up a small component of total systems in marine environments [20, 28].

2.2 Coelenterazine

The majority of luminescent organisms in marine environments with known or partially studied light emission systems utilise coelenterazine. Coelenterazine is an imidazopyrazinone compound (3,7-dihydroimidazopyrazin-3-one structure) that occurs exclusively in marine organisms in a wider range of phyla (at least nine) than any other luciferin [4]. These include radiolarians, ctenophores, cnidarians, molluscs, multiple arthropods, and some fish [29]. A large proportion of these organisms are assumed to have taken up this luciferin through their diet with only a few organisms shown to synthesise their own substrate [30–32]. The coelenterazine molecule was originally given its name due to the initial discovery of its presence in coelenterates, namely *A. victoria* and *Renilla reniformis* [33]. *A. victoria* is a hydrozoan jellyfish that emits a green light at 508 nm from a ring of photocytes on the peripheral regions of its umbrella [3]. Variants of this substrate exist in several species of squid either as a coelenterazine disulphate [34] or as dehydrocoelenterazine [35, 36].

Whilst coelenterazine has been found in a diverse array of phyla, a biosynthetic pathway and origin has not yet been determined for the majority of species, which are thought to obtain coelenterazine through their diet [12]. Coelenterazine has been shown to be synthesised in the deep-sea copepod, *Metridia pacifica,* via a similar mechanism to that observed for cypridinid luciferin in *Vargula hilgendorfii* wherein free amino acids are biosynthesised to form the coelenterazine luciferin [20, 26]. By labelling L-tyrosine and L-phenylalanine with deuterium it was proven that *M. pacifica* was able to incorporate these amino acids into its diet and that it was able to synthesise coelenterazine from two molecules of L-tyrosine and one molecule of L-phenylalanine [14]. Given that *M. pacifica* is at a lower trophic level it is likely to be predated upon by several higher taxa, many of which exhibit their own luminescent reactions [14, 37].

Recently it has been proposed that luminescent ctenophores are also able to produce their own luminescent components. The phylum Ctenophora or comb jellies are similar to the coelenterates in their morphology and apart from the family Pleurobrachiidae, all are presumed to be luminescent [38]. Ctenophores had previously been considered to be a source of coelenterazine synthesis in the oceans as there are reports of bioluminescence at early developmental stages [39]. When fed a coelenterazine-free non-luminescent diet, ctenophores were still shown to possess this substrate via mass spectrometry [40]. This recent study has implications that a number of other marine organisms, in addition to *M. pacifica* and Ctenophora, have the capacity to synthesise luciferin, which can provide a clear source of coelenterazine for a number of semi-intrinsic luminescent organisms.

3. Semi-intrinsic luminescent systems

3.1 Luminescence in fish

Most notable semi-intrinsic luminescence occurs in higher trophic levels such as among fishes. Several species have been shown to utilise the imidazopyrazinone type substrates cypridinid luciferin and coelenterazine in luminescent reactions [1, 3, 41], though they are shown to express their own luciferase enzymes [6]. Often these have evolved to harbour luminescence in specialised regions of the body that allow for particular behaviours and functions for luminescence [1, 42].

3.1.1 Cypridinid luciferin in the midshipman fish

Several species of midshipman fish have been shown to utilise cypridinid luciferin as a substrate in their own luminescent reactions, despite showing no identifiable capability to synthesise their own luciferin [43]. A notable example of this has been observed consistently in the species *Porichthys notatus*, which can be found along the Pacific coast of the North American continent [44]. This species is characterised by an array of over 700 dermal photophores distributed along its head and body [45, 46]. Whilst light emission is restricted to specific organelle structures and can be stimulated mechanically, this is not sufficient to constitute a wholly intrinsic luminescent system. Moreover, non-luminescent individuals of the species have been identified when caught in the North Pacific off the coast of Oregon, where despite possessing the photophores in the same pattern, they did not exhibit luminescence [47]. This lack of luminescence was attributed to these animals not having a source of luciferin available from their diet at all of their life stages [48].

By adding small amounts of cypridinid luciferin to *P. notatus*, either by feeding them ostracods, or by intraperitoneal doses of as little as 6 μg of luciferin it was possible to induce luminescence [44]. This also was shown to be possible for completely non-luminescent individual midshipman fish and confirmed cross-reactivity of *P. notatus'* luciferase with cypridinid luciferin led to light emission [43]. It was identified that following consumption of ostracods, *P. notatus* is able to absorb the cypridinid luciferin through its gut. From here the substrate is believed to be able to bind non-specifically to erythrocytes in the blood plasma, possibly preventing autooxidation as it is transferred to the organelles of *P. notatus* where it can be oxidised in the presence of the luciferase enzyme to result in an emission of blue light [43, 49]. Light emission from the addition cypridinid luciferin to non-luminescent *P. notatus*, was indistinguishable from naturally luminescent Californian *P. notatus* [49].

The midshipman fish is a visually active nocturnal predator, that can utilise this acquired cypridinid luciferin to facilitate its hunting strategies. It has been speculated that the array of photophores on its body can mimic the light emission seen in euphausiid swarms, attracting unsuspecting prey [43, 50, 51]. This ability in combination with its highly evolved eyesight have allowed for it to be an effective nocturnal predator, feeding on both luminescent and non-luminescent organisms [52]. Cypridinid luciferin is not isolated to this species and has been found in several other luminescent coastal fishes including in the families, Pempheridae and Apogonidae [53]. Apogonids, or cardinalfishes are mostly reef dwelling with several species exhibiting visceral light organs that produce luminescence [54]. Similarly, Pempheridae commonly known as sweeper fishes, also have photophores along the length of their bodies and tend to be found in shallow marine and brackish waters [54]. It is likely that these species acquire their luminescence from ostracods, in a similar manner to the midshipman fish, though this is still to be confirmed experimentally.

3.1.2 Coelenterazine in Myctophid and Stomiid fishes

Cypridinid luciferin accounts for the luminescence observed in only a few species of bony fish as well as within ostracods, meaning it does not encompass a large amount of the total luminescence in marine environments. The most ubiquitous luciferin found in marine organisms is coelenterazine with species across multiple phyla utilising this as their substrate for light emission [12, 40]. Among the fishes, numerous species of Myctophidae and Stomiiformes have been shown to utilise coelenterazine for bioluminescence, which is obtained through their diet, either by predating directly on coelenterazine producing copepods such as *Metridia pacifica*, or indirectly by predating on the consumers of these copepods [55, 56].

Myctophids, commonly known as lanternfish, are one of the most widespread and abundant families of mesopelagic fish in the oceans. They are distributed globally, with over 250 species identified across 33 genera and 2 subfamilies [56, 57]. Lanternfish are taxonomically distinguished by specific patterns of luminescent photophores that have allowed for a diverse array of strategies for both prey detection and predator avoidance [58, 59]. Generally, Lanternfish have two kinds of photophores, one along the body with the other proximal to their eyes (**Figure 2**). These two sets of photophores are able to illuminate independently from one another allowing for a variety of ecological functions. Photophores arranged on the ventral surface produce a constant dim blue luminescent glow and can allow for counter-illumination similar to other luminescent fishes, which would allow lanternfish to blend into the surrounding water column [56]. This would facilitate an ability to ambush prey as well as to hide from potential predators in the water column. These arrays of photophores form species specific patterns, which may allow for them to be used in intraspecific recognition [56, 60]. In addition to this array of photophores on the body, most lanternfish have one or more larger photophores on their head, usually positioned sub-orbitally or in the direct vicinity of their eyes [61]. Unlike the photophores on the ventral surface, these emit light in brief intermittent brilliant flashes. This is thought to allow either for predation by illuminating their prey, as well as being used to avoid predators by flashing and startling any larger organisms [56, 62]. Given that these suborbital photophores have sexual dimorphism, it is also possible that their main role is in communication within the species [56, 63].

Lanternfish feed predominantly on a variety of zooplankton including copepods such as *M. pacifica*, which would facilitate a source of coelenterazine luciferin for their luminescence, although it is difficult to assess this given the difficulties of maintaining deep sea fish such as myctophids in aquaria for sufficient amounts of time [55]. Lanternfishes are a major food source for a number of marine predators, including whales and dolphins. More importantly, they are also predated upon by squid and other larger lanternfishes, that also possess luminescence using coelenterazine or one of its derivatives [59]. Therefore, these potentially provide a key link in food webs by facilitating the transfer of coelenterazine from zooplankton to megafauna.

Stomiiform fishes include four families comprising of Gonostomatidae (bristlemouths), Phosichthyidae (lightfishes), Sternoptychidae (hatchetfishes), and the Stomiidae (dragonfishes) [64]. Among the dragonfishes, all species identified within this group have been shown to be bioluminescent, harbouring their light emission within specialised arrays of photophores. Apart from the Arctic Ocean, Stomiidae fishes are distributed globally, residing in the mesopelagic zone of the ocean between 200 and 1000 m depth, with some species recorded to a depth more than 4000 m [64, 65]. Luminescence may well be derived from the coelenterazine in their diets, with several species showing cross reactivity with coelenterazine in a similar way to some lanternfish [3]. However, it has been difficult to determine whether these animals are capable of synthesising their own luciferin, given that it is not yet possible to collect and maintain stomiid fishes in aquaria for any length of time. Dragonfishes are predators, utilising their bioluminescent emissions both as lures and as means to illuminate prey in order to facilitate prey capture [64]. Most feed on squid, shrimps and other fishes including lanternfishes, which may facilitate a source for coelenterazine in a number of these species [64].

Figure 2.
Photographs of Diaphus *sp. captured from a lateral (upper) and ventral view (middle). Displaying the photophores that produce a blue luminescent light (lower). Photographs taken by Yuichi Oba.*

Support for a dietary origin for luciferin in a number of stomiids is supported by their ability to uptake other small molecules to utilise in light emission. An example of this is shown in several species of loose-jaw dragonfish (*Malacosteus* spp.), that have a rare ability to emit longer wavelengths of luminescence that is red in colour, as opposed to blue light which is more ubiquitous in the oceans [1]. *Malacosteus* can also detect red wavelengths of light using a distinct mechanism requiring derivatives of bacteriochlorophylls *c* and *d* that enhance its sensitivity to these longer wavelengths [66]. As vertebrates are unable to synthesise chlorophyll, *Malacosteus* could obtain this through a diet, predominantly of grazers such as copepods that will contain phytoplankton derived pigments in their guts [64]. This strongly supports the concept that other small organic compounds such as luciferins can be taken up by dragonfishes, as well as other Stomiiformes to utilise in their bioluminescent reactions.

3.2 Other Coelenterazine utilising systems

Semi-intrinsic luminescence is clearly present in several marine vertebrates that utilise either cypridinid luciferin or coelenterazine as their substrate. However, this alone does not account for the diverse array of marine phyla that use coelenterazine in their bioluminescent behaviours. Many organisms previously considered to synthesise coelenterazine have since been shown to obtain this through their diet, including in the cnidarians where this was first discovered.

3.2.1 Cnidaria (Coelenterates)

Bioluminescence within the phylum Cnidaria has been studied more than in any other marine invertebrate. Most notably the hydromedusa *A. victoria* which emit light via the enzymatic oxidation of coelenterazine in the presence of calcium [12]. Unlike most coelenterazine utilising organisms that emit blue light, in *A. victoria*, light emission is green due to a green fluorescent protein. This emits green light via resonance energy transfer from the aequorin photoprotein [67]. According to Shimomura [3], photoproteins can be distinguished from luciferases by two general means, not requiring molecular oxygen for light emission and being capable of emitting light proportional to the amount of protein present [68]. Isolated aequorin can appear to emit light only by adding Ca^{2+}, and once the reaction is complete the protein does not appear to immediately be available for further reactions [69].

By controlling the diet of *A. victoria* in the lab it was possible to show that they are dependent on a dietary supply for their luciferin. When provided with an external source of luciferin to uptake after this, *A. victoria* was able to regain its luminescence [12]. The diet of *A. victoria* will consist of a variety of zooplankton, including luminescent copepods such as *M. longa* as well as luminescent ctenophores, which could provide a dietary source for their luminescence. Several other notable examples of luminescent coelenterates are presumed to obtain coelenterazine from their diet including the sea pansy, *Renilla* sp. and the sea cactus *Cavernularia obesa* [70, 71]. These anthozoans are found predominantly in tropical waters and may be able to obtain coelenterazine by feeding on suspended detrital matter that may contain the substrate.

3.2.2 Crustacea

Among the crustacea there is proven case of a fully intrinsic luminescent system in the copepod *Metridia pacifica*, and a probable case in the decapod shrimp *Systellaspis debilis* which appears to have the ability to synthesise the molecule from free amino acids [72]. Zooplanktonic species such as these potentially provide a source for a lot of the coelenterazine utilised in semi-intrinsic luminescent systems found in many marine organisms. However not all crustacea are able to perform this, and some such as the lophogastrid shrimp, *Neognathophausia ingens*, have been shown to require coelenterazine from their diet [31, 73].

These shrimp use bioluminescence to evade predators as they emit a brilliant blue cloud of luminescence when agitated that acts as a smoke screen [74]. Given that deep water visual predators have highly sensitive eyes, the bioluminescent ink cloud will have a much greater effect in startling nearby predators than the ink clouds produced by most cephalopods [75]. It is possible that producing this amount of luminescent material has a high energetic so it may be easier from an evolutionary perspective to obtain this through their diet instead of via an internal biosynthetic pathway.

3.2.3 Radiolaria

An assumption may be that as the majority of coelenterazine in the ocean is produced and utilised by eukaryotes, that organisms such as protists would synthesise their own source of luciferin rather than obtain it through their diets. However even protozoa such as several radiolarian species are not only capable of bioluminescence but obtain coelenterazine through their diet [1]. For example, bioluminescence has been found in several species of *Thalassicolla* and *Sphaerozoum* [29]. As protists they may appear to be unable to possess semi-intrinsic luminescence, however these species are heterotrophic, and capable of consuming and digesting larger prey including zooplanktonic copepods [76]. As to the function of luminescence in these organisms it remains poorly understood, although given their dietary acquisition of luciferin, light emission may assist in prey attraction and capture [1].

3.2.4 Chaetognatha

Other smaller marine organisms are able to acquire luminescence through predation, such as at least two species of chaetognaths. This phylum comprises of small, elongated worms that are between 2 and 120 mm in length [77]. Commonly known as "arrow worms" at least two species have been shown to be luminescent and can be found at depths greater than 700 m in marine systems ranging from tropical to polar regions [78]. Luminescence in all of these species is emitted as a blue cloud of light and may facilitate a role in stunning their prey to assist with their hunting strategies giving the lack of visible light that will attenuate down to these depths. Despite being from evolutionarily distinct lineages within the chaetognaths, luminescent species such as *Caecosagitta macrocephala* [79] and *Eukrohnia fowleri*, have a relatively uncommon trait among chaetognaths, in that they have an orange-pigmented gut lining [80]. Digestive systems in semi-transparent organisms that are orange in colour, have the capacity to mask any luminescence produced by ingested prey [78].

This provides strong evidence that some species will predate on luminescent organisms such as copepods in order to provide a dietary source of coelenterazine for their luminescent reaction as shown in a number of other marine organisms [12, 48]. Once absorbed, coelenterazine would be able to be passed through to their luminescent organs that harbour the light reaction, which tend to be found on the lateral and dorsal fins as well as along the sides of the body of these species [78].

3.2.5 Ophiuroidea

Most species that exhibit semi-intrinsic bioluminescence acquire their luciferin via predation, most notably on luminescent copepods or on their predators. However, it is also feasible that filter feeders will be able to acquire coelenterazine and other luciferins through their diet. One such example is seen in the ophiuroids or brittle stars where many species have been shown to emit light [81, 82]. One such example is the brittle star *A. filiformis*, whose bioluminescence has been studied from a biochemical perspective for the past decade. This species feeds on suspended organic matter by extending its arms into the water column [83, 84]. Each of its arms are covered with light-emitting cells called photocytes that have been shown to be dependent on coelenterazine as a source of luciferin [81, 84, 85]. Additionally, the enzyme involved in its luminescent reaction was shown to be homologous to *Renilla* luciferase, which is a coelenterate also thought to acquire its luciferin from its diet [81, 86].

A recent study monitored *A. filiformis* kept in an aquarium for several months whilst controlling its diet [82]. Over five months a depletion in *A. filiformis'* luminescence was observed when fed a coelenterazine-free diet, strongly suggesting it acquired components for luminescence through filter feeding [82]. This was validated as there was a quick recovery in its luminescent capabilities once the brittle star was fed coelenterazine supplemented food. This animal signifies that semi-intrinsic luminescent systems are not simply found among tertiary consumers. This also supports the notion that numerous other filter and detrital feeding organisms that exhibit luminescence, acquire their substrates via their diet.

3.2.6 Tunicata

While it has not fully been confirmed yet, it is possible a large number of other filter feeding marine organism can acquire luminescent components from their diet. Within the chordates the subphylum Tunicata, comprises of a number of species shown to produce luminescence, although compared with other luminescent organisms these remain poorly studied. Within the tunicates, luminescence is well represented among the appendicularians with several species being confirmed to produce luminescence. One such example within this group is the larvacean *O. dioica*, which is a free-swimming tunicate that dwells in the photic zone of the ocean [87]. The animal has transparent body and a tadpole-like appearance throughout its life cycle, ranging in size from 0.5 to 1 mm. Light emission occurs as blue flashes of light from its body that can be induced by mechanical stimulation [88]. This animal has also been reported to emit light in the presence of coelenterazine, so it is possible that these are able to acquire coelenterazine from exogenous sources [87]. Larvaceans like *O. dioica* can secrete their luminescence as a mucus that will capture and collect particulate organic matter whilst the animals are filter feeding [89]. These secretions form luminescent "houses" or clusters of organic matter which can harbour all of the components for the bioluminescent reaction. On mechanical stimulation, these "houses" emit blue light showing that the components luminescence are all present in a way such that coelenterazine does not undergo autooxidation. This display of luminescence supports coelenterazine being utilised by this and other filter feeders for semi-intrinsic luminescence as stable luciferins can potentially be found in particulate organic matter that these organisms can feed on [87, 88, 90].

Another example of luminescence in tunicates is found in pyrosomes which are pelagic tunicates known for their sustained bright blue luminescence as well as their capacity to form sporadic and yet massive blooms such as those observed in this region [91]. There is currently a lack of consensus on the origin of luminescence in this species. A recent study has shown that light emission occurs in the presence of coelenterazine for the species *Pyrosoma atlanticum* [92]. Moreover, using transcriptomic analysis, an enzymatic sequence was identified as being similar to the luciferase found in the Cnidarian *Renilla reniformis* that uses coelenterazine as its light emitter. Subsequent expression of this gene showed that light emission occurred in the presence of coelenterazine strongly supporting that this is the luciferase involved in pyrosome bioluminescence [92]. Coelenterates and some echinoderms have been shown to utilise luciferases with a similar structure to *Renilla*, and a number of these are thought to acquire coelenterazine through their diets. Therefore, it is entirely feasible that pyrosomes such as this species attain coelenterazine through filter feeding, which may also occur for various other luminescent tunicates. However, it should also be noted that recent studies have identified and characterised potentially luminescent bacterial symbionts within *P. atlanticum* [93] which supports several previous studies on this system. Determining how this organism obtains its luminescence will rely on further confirmation what the source of light emission is in this tunicate.

3.2.7 Mollusca

Like previously mentioned phyla, some luminescent molluscs are able to acquire coelenterazine through their diet. This includes the clam *Pholas dactylus*, as well as several species of squid that have been shown to possess coelenterazine in their livers [94]. However, these animals do not use coelenterazine directly as their source of luciferin for bioluminescence. Instead, they use modified forms of this substrate, for example the firefly squid utilises a disulphate form of coelenterazine

Figure 3.
Photograph of Watasenia scintillans *taken under natural light (upper) and in a dark room (lower) showing the luminescent photophores along its body. Photographs taken by Yuichi Oba.*

in its luminescent reaction [95]. These produce a dim continuous blue bioluminescence from ventral photophores, as well as a bright blue flash of luminescence (470 nm) from light organs on its arm tips after being mechanically stimulated [96]. The flashing ability may be used as a means of intra-specific communication and recognition although this has not yet been defined. The enzymatic oxidation of coelenterazine disulphate [luciferin] in the presence of Mg^{2+} has led to emissions of blue light, however how or why obtained coelenterazine is modified remains undetermined [95, 97].

Another derivative found in several molluscs is dehydrocoelenterazine. This is an oxidised form of coelenterazine and was identified as the luciferin required in the luminescence of the clam *P. dactylus*, the purple back flying squid *Sthenoteuthis oualaniensis* and recently the Humboldt squid *Dosidicus gigas* [98]. In *D. gigas*, a blue bioluminescent light is emitted from an array of photophores on their body [39]. These structures are small, ovoid rice-like granules that are embedded in the muscle all over the squid on the mantle, fins, head, arms and tentacles [99]. It is entirely possible that this and other squids can obtain coelenterazine from lanternfishes which they are known to predate on. This coelenterazine may undergo an enzymatic oxidation to form dehydrocoelenterazine which is then utilised in its light emission (**Figure 3**).

3.3 Non imidazopyrazinone substrates

All examples of semi-intrinsic luminescence so far have involved either coelenterazine or cypridinid luciferin as the substrate. Dinoflagellate luciferin has also been shown be required by several heterotrophic organisms that appear to not be able to synthesise this luciferin. Dinoflagellates are unicellular organisms that account for the majority of bioluminescence observed in the surface ocean [100, 101]. The compounds involved with luminescence are regulated on a diurnal circadian rhythm, along with photosynthetic components. This means that dinoflagellates conduct primary production during the day and only produce bioluminescence at night, when this would be most effective. The structure of this luciferin was originally determined from *Pyrocystis lunula*. The compound is a linear tetrapyrrole which is very sensitive to non-enzymatic oxidation and is most likely to have derived from chlorophyll [102]. Within different species of dinoflagellates there is variation in the intensity and duration of light emission but in general light is emitted from organelles known as scintillons [101].

Dinoflagellate luciferin shows no similarities to other luciferins and is found in forms, one within dinoflagellates and another with two hydroxyl moieties in euphausiids (krill). This similarity suggests that there is some form of dietary link [102, 103]. Studies have shown luminescent euphausiids occurred in high densities which coincided with large populations of dinoflagellates during late spring [104]. Additionally, heterotrophic species of dinoflagellate, such as *Noctiluca scintillans* have been shown to feed on luminescent dinoflagellates such as *P. lunula*. When their diet was controlled in the lab to exclude luminescent dinoflagellates and all other phytoplankton, they were shown to lose their capacity to emit light [101]. Moreover, when fed other non-dinoflagellate phytoplankton, luminescence was maintained, suggesting that *N. scintillans* can synthesise the tetrapyrrole luciferin from chlorophyll [105]. These examples suggest other luciferins and their precursors may be taken up in the diets and utilised by consumers that already express the required luciferases for other non-imidazopyrazinone luciferins.

4. "Kleptoprotein" luminescence

A general consensus among semi-intrinsic luminescent systems is that the components of the light emission utilised by other organisms are the substrates rather than enzymes. As most of these animals acquire luminescence through their diets, any exogenous components would need to be able to withstand digestion and potentially transport through the blood plasma to the luminogenic organs. Given this it seems unlikely that the enzymatic component of luminescence would be able to be obtained in this manner, as they would likely be denatured and completely broken-down during digestion [13].

However, a recent study on the *Parapriacanthus* fish, has shown that it is able to obtain both its luciferin and luciferase from its prey. Like midshipman fish, *Parapriacanthus ransonneti* predates on ostracods, which provide a source of cypridinid luciferin that is used in its light emission [13]. When *P. ransonneti* was fed on the ostracod *Cypridina noctiluca,* the luciferase identified from its light organs was identical to the luciferase of this species. When a different species of luminescent ostracod, *Vargula hilgendorfii* was identified in another individual fish, the identified luciferase was now the same as this ostracod, demonstrating the ability to specifically uptake luciferases from its diet to the fish's light organs [13]. Transcriptomic analysis of *P. ransonneti*, showed no transcripts corresponding to an ostracod-type luciferase, further highlighting that this was acquired via the diet (**Figure 4**).

This is the first reporting of this type of phenomenon in bioluminescence, and up until now it was assumed that any consumed luciferase enzyme would be broken down into amino acids or oligopeptides before being absorbed via the gut wall as nutrients [13]. However, the possibility of protein uptake without being fully broken down and retaining activity has been reported in several vertebrate immune systems. An example of this is seen in M cells within the mammalian intestinal epithelia as these have an important role in the immune system by transporting macromolecules and microbes into the cell via pinocytosis [106]. Similar examples of this have been observed in cyprinid fishes so it is feasible these or similar structures could facilitate the transfer of ostracod luciferase to the photophores of this animal [13].

This example of a "kleptoprotein" form of luminescence where both the substrate and the enzyme are provided through the diet, provides an additional novel category of luminescent reactions, as of yet not considered. Moreover, this highlights the possibility that other luminescent species may utilise this capability to obtain active exogenous luciferase from their gut. Potentially, this may include several species of fishes that predate on ostracods, whose light organs are often connected to their digestive tracts. This research may suggest that semi-intrinsic and "kleptoprotein" luminescent behaviours may be more widespread than previously considered, with proteins associated with other biological processes potentially being able to be attained via diet as opposed to gene expression.

Figure 4.
Ventral view of Parapriacanthus ransonneti *taken in a dark room to capture the light emission from these body regions. Photos by Okinawa Commemorative National Government Park (Ocean Expo Park), Okinawa Churaumi Aquarium.*

5. Why semi-intrinsic luminescence occurs?

Semi-intrinsic luminescence has been shown to exist in a number of organisms and is hypothesised to exist in several others. Cypridinid luciferin and dinoflagellate luciferin have been shown to be taken up by predators of ostracods and dinoflagellates respectively, notably several species of fishes, and euphausiid shrimp. However, the majority of semi-intrinsic luminescence, in addition to the majority of bioluminescence in the oceans involves using coelenterazine. Dietary uptake of coelenterazine has been shown in coelenterates, echinoderms, and decapod shrimp, while it is also strongly supported to be the source of luciferin in myctophid and stomiid fishes, chaetognaths, tunicates and several species of squid. Moreover, coelenterazine can be modified via oxidation or di-sulfonation, once it is taken up by species, allowing for a variety of different light reaction mechanisms to occur with this molecule. It is important to understand why some animals use semi-intrinsic luminescence, and the potential evolutionary origins of this, and how coelenterazine may spread across the food web and be the most common light emission system in the oceans. It is useful to consider whether this phenomenon along with "kleptoprotein" luminescence is a lot more widespread in other biological processes and systems.

There are two main groups of hypotheses on why bioluminescence evolved originally; one based around changes in the luciferin (substrate-centric hypothesis) [5, 107] and another that suggests changes occurred in what became the luciferase enzyme (enzyme-centric hypothesis) [108]. The first hypothesis suggests that the luciferin substrate evolved in order to protect organisms from reactive oxidative species (*e.g.*, hydrogen peroxide) in the water column [108]. Luminescent animal migrated to deeper water to evade visual predators and at these depths there was no longer significant oxidative stress. Therefore, the active selection pressure switched to the luminescent, communicative properties of luciferins, leading to more specific adaptations to predation, survival, and communication [1].

The alternate hypothesis focuses on the enzyme luciferase and that these molecules were originally less specific oxygenase enzymes [108]. The oxygenase enzymes mutated as a result of animals migrating to deeper waters to either evade visual predators, or to predate on organisms that have migrated to deeper water [5]. The mutation in oxygenase enzymes associated with display functions would result in external luminescence being exhibited [109]. These display pigments would previously have been associated with warning colourations or patterns to both recognise species and attract potential mates. There is evidence for enzyme-based hypotheses in terms of enhancement of visual signals [5]. However, there is no biochemical or genetic evidence that would support this hypothesis, and the mutation of the luciferase enzyme alone would not explain the convergent evolution of the bioluminescent reaction in multiple phyla [1, 5].

Whether one or a combination of both hypotheses are more viable for the origins of luminescence, both allow for the possible co-evolution of predators and prey that may utilise the same source of luminescence. Convergent evolution caused by environmental factors may have allowed for the presence of various enzymes that were compatible with the same substrate resulting in coelenterazine being utilised by both animals that can synthesise it as well as their predators. Moreover, given the energetic costs associated with synthesising luciferins, it may simply be more efficient for some of these organisms to acquire exogenous sources instead.

Semi-intrinsic luminescent organisms, particularly those that harbour coelenterazine, have shown the potential spread and dispersal across the food web for not just luciferins, but other molecules that may be involved in biological processes. A major source of coelenterazine is found in the copepod *M. pacifica* which is grazed upon by a variety of organisms including coelenterates, lanternfishes, euphausiids, and radiolarians. Additionally, these animals, particularly lanternfishes are predated upon by tertiary consumers such as squid, stomiid fishes and luminescent sharks [110]. The consumption of copepods by zooplankton and higher taxa can lead to particulate organic matter or marine snow forming and descending to the depths of the ocean. These aggregates will contain detritus, plankton and larvacean houses, meaning that it is highly likely for free-available coelenterazine to be present. The coelenterazine within this particulate organic matter can then be taken up by filter feeders such as echinoderms and tunicates, allowing for them to utilise coelenterazine in their luminescent displays.

In a number of these organisms, luciferin has been identified in a sulfonated form. The most notable example of this is in the firefly squid, however sulfonated luciferins have been identified in *V. hilgendorfii* and *Renilla reniformis* [3]. This form is more stable than free forms of coelenterazine, and it is possible this is a stored form of luciferin that may prevent auto-oxidation that can occur. This more stable form may prevent breakdown and oxidation of the substrate when it is in the water column or during digestion. Potentially, a lot of these semi-intrinsic luminescent organisms will obtain their luciferins in this form, and then have the capability to de-sulfonate the luciferin to make it available for luminescence.

6. Conclusions

Luminescence has evolved and been prevalent in a wide variety of marine species being utilised for a number of predative, defensive and communicative functions. Some organisms have developed predator–prey relationships where the predator is able to acquire and utilise luciferin with its own luciferase to emit light. This chapter has reviewed many of the species that exhibit this type of behaviour and utilise semi-intrinsic luminescence, in addition to describing the sources of luciferin in these systems and how this molecule is able to be taken up by consumers. Although this has only been experimentally tested in a few species, it is highly likely that a number of other luminescent organisms utilise this, especially as it is a lot easier from an evolutionary perspective to obtain luciferins from the diet, compared with synthesising them from amino acids or other unknown biosynthetic pathways. This phenomenon raises the question of whether small molecules and enzymes involved in other biological processes are able to be taken up in this manner as well which could provide an evolutionary selection process that is an alternative to molecular evolution.

Notes/thanks/other declarations

This work is supported in part by JST CREST (JPMJCR16N1). We would also like to thank Dr. Ken-ichi Onodera and Okinawa Churaumi Aquarium for providing some photos for this chapter.

Author details

Jeremy Mirza[1] and Yuichi Oba[2*]

1 University of Birmingham, United Kingdom

2 Chubu University, Kasugai, Japan

*Address all correspondence to: yoba@isc.chubu.ac.jp

References

[1] Haddock SHD, Moline MA, Case JF. Bioluminescence in the sea. Annual review of Marine Science. 2010;15:443-493. DOI: 10.1146/annurev-marine-120308-081028

[2] Wilson T, Hastings JW. Bioluminescence. Annual Review of Cell and Developmental Biology. 1998;14:197-230. DOI:10.1146/annurev.cellbio.14.1.197

[3] Shimomura O, Yampolsky I, editors. Bioluminescence: chemical principles and methods. 3rd ed. Singapore: World Scientific; 2019. DOI: 10.1142/11180

[4] Lau ES, Oakley TH. Multi-level convergence of complex traits and the evolution of bioluminescence. Biological Reviews. 2021;96:673-91. DOI:10.1111/brv.12672

[5] Widder EA. Bioluminescence in the ocean: origins of biological, chemical, and ecological diversity. Science. 2010;328:704-708. DOI: 10.1126/science.1174269

[6] Davis MP, Sparks JS, Smith WL. Repeated and widespread evolution of bioluminescence in marine fishes. PloS ONE. 2016;11:e0155154. DOI:10.1371/journal.pone.0155154

[7] Pieribone V, Gruber DF. Aglow in the dark: the revolutionary science of biofluorescence. Cambridge (Massachusetts): Harvard University Press; 2005. 263 p DOI: 10.1086/511547

[8] Haddock SHD. Luminous marine organisms. In: Daunert S, Deo SK, editors. Photoproteins in Bioanalysis. Wenheim: Wiley-VCH; 2006. p. 25-47. DOI: 10.1002/3527609148

[9] Martini S, Haddock SHD. Quantification of bioluminescence from the surface to the deep sea demonstrates its predominance as an ecological trait. Scientific Reports. 2017;7:45750. DOI: 10.1038/srep45750

[10] Kaskova ZM, Tsarkova AS, Yampolsky IV. 1001 lights: luciferins, luciferases, their mechanisms of action and applications in chemical analysis, biology, and medicine. Chemical Society Reviews. 2016;45:6048-6077. DOI: 10.1039/C6CS00296J

[11] Martini S, Schultz DT, Lundsten L, Haddock SHD. Bioluminescence in an undescribed species of carnivorous sponge (Cladorhizidae) from the deep sea. Frontiers in Marine Science. 2020;7:1041. DOI: 10.3389/fmars.2020.576476

[12] Haddock SHD, Rivers TJ, Robison BH. Can coelenterates make coelenterazine? Dietary requirement for luciferin in cnidarian bioluminescence. Proceedings of the National Academy of Sciences of the United States of America. 2015;98:11148-11151. DOI: 10.1073/pnas. 201329798

[13] Bessho-Uehara M, Yamamoto N, Shigenobu S, Mori H, Kuwata K, Oba Y. Kleptoprotein bioluminescence: *Parapriacanthus* fish obtain luciferase from ostracod prey. Science Advances. 2020;6:eaax4942. DOI:10.1126/sciadv.aax4942

[14] Oba Y, Kato S, Ojika M, Inouye S. Biosynthesis of coelenterazine in the deep-sea copepod, *Metridia pacifica*. Biochemical and Biophysical Research Communications. 2009;390:684-8. DOI: 10.1016/j.bbrc.2009.10.028

[15] Dubois, R. Note sur la physiologie des Pyrophores. Comptes rendus des séances de la Société de biologie, Series 8. 1884;1:661-664.

[16] Harvey, EN Studies on Bioluminescence. XIV. The specificity of luciferin and luciferase. The Journal of General Physiology. 1922;4:285-295.

[17] Wilson T, Hastings JW. Bioluminescence: living lights, lights for living.

Cambridge (Massachusetts): Harvard University Press; 2012. 176 p DOI: 10.4159/harvard.9780674068025

[18] Viviani VR, Bechara EJH, Ohmiya Y. Cloning, sequence analysis, and expression of active *Phrixothrix* railroad-worms luciferases: relationship between bioluminescence spectra and primary structures. Biochemistry. 1999;38:8271-8279. DOI: 10.1021/bi9900830

[19] Thomson CM, Herring PJ, Campbell AK. The widespread occurrence and tissue distribution of the imidazolopyrazine luciferins. Journal of Bioluminescence and Chemiluminescence. 1997;12:87-91. DOI: 10.1002/(SICI)1099-1271 (199703/04)12:2<87::AID-BIO438>3.0.CO;2-8

[20] Oba Y, Kato S, Ojika M, Inouye S. Biosynthesis of luciferin in the sea firefly, *C. hilgendorfii*: L-tryptophan is a component in *Cypridina* luciferin. Tetrahedron Letters. 2002;43:2389-92. DOI:10.1016/S0040-4039(02)00257-5

[21] Shimomura O, Goto T, Hirata Y. Crystalline *Cypridina* luciferin. Bulletin of the Chemical Society of Japan. 1957;30:929-933. DOI: 10.1246/bcsj.30.929

[22] Tsuji FI. The absorption spectrum of reduced and oxidized *Cypridina* luciferin, isolated by a new method. Archives of Biochemistry and Biophysics. 1955;59:452-464. DOI: 10.1016/0003-9861(55)90511-7

[23] Kishi T, Goto T, Hirata Y, Shimomura O, Johnson FH. *Cypridina* bioluminescence I Structure of *Cypridina* luciferin. Tetrahedron Letters. 1966;7:3427-3436. DOI: 10.1016/S0040-4039(01)82806-9

[24] Shimomura O, Johnson FH, Masugi T. *Cypridina* bioluminescence: light-emitting oxyluciferin-luciferase complex. Science. 1969;164:1299-1300. DOI: 10.1126/science.164.3885.1299

[25] Shimomura O, Johnson FH. Mechanisms in the quantum yield of *Cypridina* bioluminescence. Photochemistry and Photobiology. 1970;12:291-295. DOI:10.1111/j.1751-1097.1970.tb06061.x

[26] Kato S, Oba Y, Ojika M, Inouye S. Biosynthesis of *Cypridina* luciferin from free amino acids in *Cypridina* (*Vargula*) *hilgendorfii*. In: Tsuji A, Maeda M, Kricka L, Stanley P, editors. Bioluminescence and Chemiluminescence: Progress and Perspectives. Singapore: World Scientific; 2005. p. 121-124. DOI: 10.1142/9789812702203_0028

[27] Kato S, Oba Y, Ojika M. Biosynthesis of *Cypridina* luciferin in *Cypridina noctiluca*. Heterocycles. 2007 13;72: 673-676.

[28] Warner JA, Case JF. The zoogeography and dietary induction of bioluminescence in the midshipman fish, *Porichthys notatus*. The Biological Bulletin. 1980;159:231-246. DOI: 10.2307/1541021

[29] Campbell AK, Herring PJ. Imidazolopyrazine bioluminescence in copepods and other marine organisms. Marine Biology. 1990;104:219-225. DOI: 10.1007/BF01313261

[30] Tsuji FI, Barnes AT, Case JF. Bioluminescence in the marine teleost, *Porichthys notatus*, and its induction in a non-luminous form by *Cypridina* (ostracod) luciferin. Nature. 1972;237: 515-516. DOI: 10.1038/237515a0

[31] Frank TM, Widder EA, Latz MI, Case JF. Dietary maintenance of bioluminescence in a deep-sea mysid. The Journal of Experimental Biology. 1984;109:385-389.

[32] Harper RD, Case JF. Disruptive counterillumination and its anti-predatory value in the plainfish midshipman *Porichthys notatus*. Marine

Biology. 1999;134:529-540. DOI: 10.1007/s002270050568

[33] Shimomura O, Johnson FH. Chemical nature of bioluminescence systems in coelenterates. Proceedings of the National Academy of Sciences of the United States of America. 1975. 72(4):1546-1549. DOI: 10.1073/pnas.72.4.1546

[34] Tsuji FI. Bioluminescence reaction catalyzed by membrane-bound luciferase in the "firefly squid," *Watasenia scintillans*. Biochimica et Biophysica Acta (BBA)-Biomembranes. 2002;1564:189-197. DOI: 10.1016/S0005-2736(02)00447-9

[35] Isobe M, Kuse M, Yasuda Y, Takahashi H. Synthesis of ^{13}C-dehydrocoelenterazine and model studies on *Symplectoteuthis* squid bioluminescence. Bioorganic & Medicinal Chemistry Letters. 1998;8:2919-2924. DOI: 10.1016/S0960-894X(98)00525-3

[36] Galeazzo GA, Mirza JD, Dorr FA, Pinto E, Stevani CV, Lohrmann KB, Oliveira AG. Characterizing the bioluminescence of the Humboldt squid, *Dosidicus gigas* (d'Orbigny, 1835): one of the largest luminescent animals in the world. Photochemistry and Photobiology. 2019;95:1179-1185. DOI: 10.1111/php.13106

[37] Tessler M, Gaffney JP, Crawford JM, Trautman E, Gujarati NA, Alatalo P, Pieribone VA, Gruber DF. Luciferin production and luciferase transcription in the bioluminescent copepod *M. lucens*. PeerJ. 2018;6:e5506. DOI: 10.7717/peerj.5506

[38] Haddock SH, Case JF. Not all ctenophores are bioluminescent: *Pleurobrachia*. The Biological Bulletin. 1995;189:356-362. DOI: 10.2307/1542153

[39] Freeman G, Reynolds GT. The development of bioluminescence in the ctenophore *Mnemiopsis leidyi*. Developmental biology. 1973;31:61-100. DOI: 10.1016/0012-1606(73)90321-7

[40] Bessho-Uehara M, Huang W, Patry WL, Browne WE, Weng JK, Haddock SHD. Evidence for de novo biosynthesis of the luminous substrate coelenterazine in ctenophores. iScience. 2020;23:101859. DOI: 10.1016/j.isci.2020.101859

[41] Shimomura O, Inoue S, Johnson FH, Haneda Y. Widespread occurrence of coelenterazine in marine bioluminescence. Comparative Biochemistry and Physiology Part B: Comparative Biochemistry. 1980;65:435-437. DOI: 10.1016/0305-0491(80)90044-9

[42] Martini S, Kuhnz L, Mallefet J, Haddock SHD. Distribution and quantification of bioluminescence as an ecological trait in the deep-sea benthos. Scientific Reports. 2019;9:14654. DOI: 10.1038/s41598-019-50961-z

[43] Mensinger AF, Case JF. Bioluminescence maintenance in juvenile *Porichthys notatus*. The Biological Bulletin. 1991;181:181-188. DOI: 10.2307/1542501

[44] Thompson EM, Nafpaktitis BG, Tsuji FI. Dietary uptake and blood transport of Vargula (crustacean) luciferin in the bioluminescent fish, *Porichthys notatus*. Comparative Biochemistry and Physiology Part A: Physiology. 1988:89(2):203-9. DOI: 10.1016/0300-9629(88)91079-1

[45] Greene CW, Greene HH. Phosphorescence of *Porichthys notatus*, the California singing fish. American Journal of Physiology-Legacy Content. 1924;70:500-506. DOI: 10.1152/ajplegacy.1924.70.3.500

[46] Strum JM. Fine structure of the dermal luminescent organs, photophores, in the fish, *Porichthys notatus*. The Anatomical Record. 1969;164:433-461. DOI: 10.1002/ar.1091640404

[47] Strum JM. Photophores of *Porichthys notatus*: ultrastructure of innervation. The Anatomical Record. 1969;164:463-477. DOI: 10.1002/ar.1091640405

[48] Tsuji FI, Barnes AT, Case JF. Bioluminescence in the marine teleost, *Porichthys notatus*, and its induction in a non-luminous form by *Cypridina* (ostracod) luciferin. Nature. 1972;237: 515-516. DOI: 10.1038/237515a0

[49] Tsuji FI, Nafpaktitis BG, Goto T, Cormier MJ, Wampler JE, Anderson JM. Spectral characteristics of the bio-luminescence induced in the marine fish, *Porichthys notatus*, by *Cypridina* (ostracod) luciferin. Molecular and Cellular Biochemistry. 1975;9:3-8. DOI: 10.1007/BF01731727

[50] Tsuji FI, Haneda Y, Lynch III RV, Sugiyama N. Luminescence cross-reactions of *Porichthys* luciferin and theories on the origin of luciferin in some shallow-water fishes. Comparative Biochemistry and Physiology Part A: Physiology. 1971;40:163-179. DOI: 10.1016/0300-9629(71)90159-9

[51] Herring PJ, Widder EA, Haddock SHD. Correlation of bioluminescence emissions with ventral photophores in the mesopelagic squid *Abralia veranyi* (Cephalopoda: Enoploteuthidae). Marine Biology. 1992;112: 293-298. DOI: 10.1007/BF00702474

[52] Mensinger AF. Ecomorphological adaptations to bioluminescence in *Porichthys notatus*. In: Luczkovich JJ, Motta PJ, Norton SF, Liem KF, editors. Ecomorphology of fishes. Dordrecht; Springer; 1995. p. 133-142. DOI: 10.1007/978-94-017-1356-6_9

[53] Paitio J, Oba Y, Meyer-Rochow VB. Bioluminescent fishes and their eyes. In: Thirumalai J, editor. Luminescence—An outlook on the phenomena and their applications. Croatia: InTechOpen; 2016. p. 297-332. DOI: 10.5772/65385

[54] Thacker CE, Roje DM. Phylogeny of cardinalfishes (Teleostei: Gobiiformes: Apogonidae) and the evolution of visceral bioluminescence. Molecular Phylogenetics and Evolution. 2009;52:735-745. DOI: 10.1016/j.ympev.2009.05.017

[55] Kinzer, J., & Schulz, K. Vertical distribution and feeding patterns of midwater fish in the central equatorial Atlantic. Marine biology, 1985;85:313-322. DOI: 10.1007/BF00393252

[56] de Busserolles F, Marshall NJ. Seeing in the deep-sea: visual adaptations in lanternfishes. Philosophical Transactions of the Royal Society B: Biological Sciences. 2017;1717:20160070. DOI: 10.1098/rstb.2016.0070

[57] Hulley PA. Myctophidae Lantern fishes. In: Carpenter KE, De Angelis N, editor. The Living Marine Resources of the Eastern Central Atlantic. Volume 3. Bony Fishes, Part 1 (Elopiformes-Scorpaeniformes), Rome: Food and Agriculture Organization of the United Nations; p. 1860-1933.

[58] Watanabe H, Moku M, Kawaguchi K, Ishimaru K, Ohno A. Diel vertical migration of myctophid fishes (Family Myctophidae) in the transitional waters of the western North Pacific. Fisheries Oceanography. 1999;8:115-127. DOI: 10.1046/j.1365-2419.1999.00103.x

[59] Karnella C. Family Myctophidae, lanternfishes. Smithsonian Contributions to Zoology. 1987;452:51-168.

[60] Edwards AS, Herring PJ. Observations on the comparative morphology and operation of the photogenic tissues of myctophid fishes. Marine Biology. 1977;41:59-70. DOI: 10.1007/BF00390582

[61] Catul V, Gauns M, Karuppasamy PK. A review on mesopelagic fishes belonging to family Myctophidae. Reviews in Fish Biology and Fisheries. 2011;21:339-354. DOI: 10.1007/s11160-010-9176-4

[62] Herring PJ. Sex with the lights on? A review of bioluminescent sexual dimorphism in the sea. Journal of the

Marine Biological Association of the United Kingdom. 2007;87:829-842. DOI: 10.1017/S0025315407056433

[63] Paxton JR. Osteology and relationships of the lanternfishes (Family Myctophidae). National History Museum of Los Angeles County. 1972;13:1-81.

[64] Sutton T. Stomiiformes (Dragonfishes and relatives). In: Thoney D, Loiselle P, editors. Grzimek's Animal Life Encyclopedia. New York: Gale; 2003. p. 421-430.

[65] Sutton T, Hopkins T. Trophic ecology of the stomiid (Pisces: Stomiidae) fish assemblage of the eastern Gulf of Mexico: strategies, selectivity and impact of a top mesopelagic predator. Mar. Biol. 1996.127(2):179-192. DOI: 10.1007/BF00942102

[66] Douglas RH, Partridge JC, Dulai KS, Hunt DM, Mullineaux CW, Hynninen PH. Enhanced retinal longwave sensitivity using a chlorophyll-derived photosensitiser in *Malacosteus niger*, a deep-sea dragon fish with far red bioluminescence. Vision research. 1999:39(17):2817-32. DOI: 10.1016/S0042-6989(98)00332-0

[67] Morise H, Shimomura O, Johnson FH, Winant J. Intermolecular energy transfer in the bioluminescent system of *Aequorea*. Biochemistry. 1974 1;13(12):2656-2662.

[68] Shimomura O. Bioluminescence in the sea: photoprotein systems. Symposia of the Society for Experimental Biology 1985. (Vol. 39, pp. 351-372).

[69] Sharpe ML, Hastings JW, Krause KL. Luciferases and Light-emitting Accessory Proteins: Structural Biology. In: eLS. Chichester: Wiley & Sons; 2014. p. 1-18. DOI: 10.1002/9780470015902.a0003064.pub2

[70] Nicol JAC. Observations on luminescence in *Renilla* (Pennatulacea). Journal of Experimental Biology. 1955;32:299-320. DOI: 10.1242/jeb.32.2.299

[71] Shimomura O, Johnson FH. Chemical nature of bioluminescence systems in coelenterates. Proceedings of the National Academy of Sciences of the United States of America. 1975;72:1546-1549. DOI: 10.1073/pnas.72.4.1546

[72] Thomson CM, Herring PJ, Campbell AK. Evidence for de novo biosynthesis of coelenterazine in the bioluminescent midwater shrimp, *Systellaspis debilis*. C. Journal of the Marine Biological Association of the United Kingdom. 1995;75:165-171. DOI: 10.1017/S0025315400015277

[73] Roe HS. Vertical migrations and feeding of mysids and decapod crustacea. Progress in Oceanography. 1984;13:269-318. DOI: 10.1016/0079-6611(84)90011-9

[74] Chan B, Lin IC, Shih TW, Chan TY. Bioluminescent emissions of the deep-water pandalid shrimp, *Heterocarpus sibogae* De Man, 1917 (Decapoda, Caridea, Pandalidae) under laboratory conditions. Crustaceana. 2008;81:341-350. DOI: 10.1163/156854008783564064

[75] Robison BH, Reisenbichler KR, Hunt JC, Haddock SHD. Light production by the arm tips of the deep-sea cephalopod *Vampyroteuthis infernalis*. The Biological Bulletin. 2003;205:102-109. DOI: 10.2307/1543231

[76] Hsin YL, Haddock SHD. The enclosing latticed sphere of *Tuscaridium cygneum* (Murray), a eurybathyal phaeodarian Radiolaria, from the North Pacific. Paleontological Research. 1997;1:144-149. DOI: 10.2517/prpsj.1.144

[77] Bone Q, Kapp H, Pierrot-Bults AC. Biology of chaetognaths. Oxford: Oxford University Press; 1991. 184 p

[78] Thuesen EV, Goetz FE, Haddock SHD. Bioluminescent organs

of two deep-sea arrow worms, *Eukrohnia fowleri* and *Caecosagitta macrocephala*, with further observations on bioluminescence in chaetognaths. The Biological Bulletin. 2010;219:100-111. DOI: 10.1086/BBLv219n2p100

[79] Haddock SHD, Case JF. A bioluminescent chaetognath. Nature. 1994;367:225-6. DOI: 10.1038/367225a0

[80] Terazaki M, Marumo R, Fujita Y. Pigments of meso-and bathypelagic chaetognaths. Marine Biology. 1977;41:119-25. DOI: 10.1007/BF00 394019

[81] Mallefet J. Echinoderm bioluminescence: where, how and why do so many ophiuroids glow? In: Meyer-Rochow VB, editor. A collection of Illuminating Essays, Research. Kerala: Singpost; 2009. p. 67-83.

[82] Mallefet J, Duchatelet L, Coubris C. Bioluminescence induction in the ophiuroid *Amphiura filiformis* (Echinodermata). Journal of Experimental Biology. 2020;223: jeb218719. DOI: 10.1242/jeb.218719

[83] Rosenberg R, Lundberg L. Photoperiodic activity pattern in the brittle star *A. filiformis*. Marine Biology. 2004;145:651-656. DOI: 10.1007/ s00227-004-1365-z

[84] Delroisse J, Ullrich-Lüter E, Blaue S, Eeckhaut I, Flammang P, Mallefet J. Fine structure of the luminous spines and luciferase detection in the brittle star *Amphiura filiformis*. Zoologischer Anzeiger. 2017;269:1-12. DOI: 10.1016/j. jcz.2017.05.001

[85] Delval S, Mallefet J. Proximal to distal gradient of luminescence in the arm of *A. filiformis* (Echinodermata-Ophiuroidea). In: Harris LG, Bottger SA, Walker CW, Lesser MP, editors. Echinoderms: Durham Proceedings of the 12th International Echinoderm Conference. New Hampshire: CRC Press; 2010. p. 355-357.

[86] Delroisse J, Ullrich-Lüter E, Blaue S, Ortega-Martinez O, Eeckhaut I, Flammang P, Mallefet J. A puzzling homology: a brittle star using a putative cnidarian-type luciferase for bioluminescence. Open biology. 2017;7:160300. DOI: 10.1098/rsob.160300

[87] Galt CP, Flood PR. Bioluminescence in the Appendicularia. In: Bone Q, editor. The Biology of Pelagic Tunicates. Oxford: Oxford University Press; 1998. p. 215-229.

[88] Galt CP, Sykes PF. Sites of bioluminescence in the appendicularians *Oikopleura dioica* and *O. labradoriensis* (Urochordata: Larvacea). Marine Biology. 1983;77:155-159. DOI: 10.1007/BF00396313

[89] Hopcroft RR, Robison BH. A new mesopelagic larvacean, *Mesochordaeus erythrocephalus*, sp. nov., from Monterey Bay, with a description of its filtering house. Journal of Plankton Research. 1999;21:1923-1937. DOI: 10.1093/plankt/ 21.10.1923

[90] Hamner WM, Robison BH. In situ observations of giant appendicularians in Monterey Bay. Deep Sea Research Part A. Oceanographic Research Papers. 1992;39:1299-1313. DOI: 10.1016/0198-0149(92)90070-A

[91] Anderson V. Salps and pyrosomid blooms and their importance in biogeochemical cycles. In: Bone Q, editor. The Biology of Pelagic Tunicates. Oxford: Oxford University Press; 1998. p. 215-229.

[92] Tessler M, Gaffney JP, Oliveira AG, Guarnaccia A, Dobi KC, Gujarati NA, Galbraith M, Mirza JD, Sparks JS, Pieribone VA, Wood RJ. A putative chordate luciferase from a cosmopolitan tunicate indicates convergent bioluminescence evolution across phyla. Scientific Reports. 2020;10:17724. DOI: 10.1109/gce.2010.5676129.

[93] Berger A, Blackwelder PL, Frank T, Sutton T, Pruzinsky N, Slayden N, Lopez JV. Microscopic and genetic characterization of bacterial symbionts with bioluminescent potential in *Pyrosoma atlanticum*. Frontiers in Marine Science. 2021:8:606818. DOI: 10.3389/fmars.2021.606818.

[94] Kuse M, Tanaka E, Nishikawa T. Pholasin luminescence is enhanced by addition of dehydrocoelenterazine. Bioorganic & Medicinal Chemistry Letters. 2008;18:5657-5659. DOI: 10.1016/j.bmcl.2008.08.113

[95] Teranishi K, Shimomura O. Bioluminescence of the arm light organs of the luminous squid *Watasenia scintillans*. Biochimica et Biophysica Acta (BBA)-General Subjects. 2008;1780:784-92. DOI: 10.1016/j.bbagen.2008.01.016

[96] Tsuji FI. ATP-dependent bioluminescence in the firefly squid, *Watasenia scintillans*. Proceedings of the National Academy of Sciences of the United states of America. 1985;82:4629-4632. DOI: 10.1073/pnas.82.14.4629

[97] Hamanaka T, Michinomae M, Seidou M, Miura K, Inoue K, Kito Y. Luciferase activity of the intracellular microcrystal of the firefly squid, *Watasenia scintillans*. FEBS Letters. 2011;585:2735-8. DOI: 10.1016/j.febslet.2011.07.033

[98] Chou CM, Tung YW, Isobe M. Molecular mechanism of *Symplectoteuthis* bioluminescence—Part 4: Chromophore exchange and oxidation of the cysteine residue. Bioorganic & Medicinal Chemistry. 2014;22:4177-4188. DOI: 10.1016/j.bmc.2014.05.044

[99] Lohrmann KB. Subcutaneous photophores in the jumbo squid *Dosidicus gigas* (d'Orbigny, 1835) (Cephalopoda: Ommastrephidae). Revista de Biología Marina y Oceanografía. 2008;43:275-284. DOI: 10.4067/S0718-19572008000200006

[100] Tett PB. The relation between dinoflagellates and the bioluminescence of sea water. Journal of the Marine Biological Association of the United Kingdom. 1971;51:183-206. DOI: 10.1017/S002531540000655X

[101] Valiadi M, Iglesias-Rodriguez D. Understanding bioluminescence in dinoflagellates—how far have we come? Microorganisms. 2013;1:3-25. DOI: 10.3390/microorganisms1010003

[102] Nakamura H, Kishi Y, Shimomura O, Morse D, Hastings JW. Structure of dinoflagellate luciferin and its enzymic and nonenzymic air-oxidation products. Journal of the American Chemical Society. 1989;111:7607-7611. DOI: 10.1021/ja00201a050

[103] Shimomura O. The roles of the two highly unstable components F and P involved in the bioluminescence of euphausiid shrimps. Journal of bioluminescence and chemiluminescence. 1995;10:91-101. DOI: 10.1002/bio.1170100205

[104] Tett PB. An annual cycle of flash induced luminescence in the euphausiid *Thysanoessa raschii*. Marine Biology. 1972.;12:207-218. DOI: 10.1007/BF00346768

[105] Buskey EJ, Strom S, Coulter C. Bioluminescence of heterotrophic dinoflagellates from Texas coastal waters. Journal of Experimental Marine Biology and Ecology. 1992;159:37-49. DOI: 10.1016/0022-0981(92)90256-A

[106] Hase K, Kawano K, Nochi T, Pontes GS, Fukuda S, Ebisawa M, Kadokura K, Tobe T, Fujimura Y, Kawano S, Yabashi A. Uptake through glycoprotein 2 of FimH+ bacteria by M cells initiates mucosal immune response. Nature. 2009;462:226-230. DOI: 10.1038/nature08529

[107] Rees JF, De Wergifosse B, Noiset O, Dubuisson M, Janssens B, Thompson EM.

The origins of marine bioluminescence: turning oxygen defence mechanisms into deep-sea communication tools. Journal of Experimental Biology. 1998;201:1211-1221. DOI: 10.1242/jeb.201.8.1211

[108] Seliger HH. Bioluminescence: Excited states under cover of darkness. Naval Research Reviews. 1993;45:5-11.

[109] Widder EA. Bioluminescence. In: Archer S, Djamgoz MB. Loew E, Partridge JC. Vallerga S, editors. Adaptive Mechanisms in the Ecology of Vision. New York: Springer; 1999. p. 555-581. DOI: 10.1007/978-94-017-0619-3_19

[110] Mizuno G, Yano D, Paitio J, Endo H, Oba Y. Lantern shark *Etmopterus* use coelenterazine as substrate for their luciferin-luciferase bioluminescence system. bioRxiv. 2021. DOI: 10.1101/2021.03.01.433353

Chapter 7

The Ecology of Bioluminescence

Aditya Srivastava and Kalpna Katiyar

Abstract

Bioluminescence, or the ability to emit light biologically, has evolved multiple times across various taxa. As fascinating as the phenomenon is, various studies have been undertaken to harness this phenomenon for human use. However, the origins, distribution and ecology of bioluminescence still remain obscure. The capability to produce biological light is found in various species, ranging from tiny bacteria to huge fishes like lantern sharks. Many organisms that do not possess this ability partake in symbiotic relationships, resulting in a variety of anatomical and behavioral modifications. The ecological interactions resulting from bioluminescence are even more interesting and diverse, but many of them are still shrouded in mystery because of a lack of *in-situ* study. As agreed by many, bioluminescence conferred certain evolutionary advantages which still remain unclear. In spite of the lack of understanding, many spectacular ecological interactions like offence, defense, courtship or intra-specific synchrony have been observed, studied and documented, and their significance understood. As far as humans are concerned, efforts are being made to channel this capability to the best of our use, though some of these are still in their infancy. This chapter explores the origins, ecology and future prospects of bioluminescence in detail.

Keywords: Bioluminescence, Ecology, Bioluminescent organisms, Firefly, Deep-sea fauna, Fungi

1. Introduction

'Bioluminescence' refers to the phenomenon of chemically induced emission of light (or other electromagnetic radiations) by a living organism. It is a common occurrence frequently observed in various organisms, ranging from simple ones like bacteria to complex animals like deep-sea fish or fireflies, and even some fungi. The first accounts of bioluminescence are found in the works of Dioscorides and even Pliny the Elder, who believed that certain bioluminescent organisms had medicinal properties [1]. There are accounts of coal-miners using dried fish skins, and even bottled fireflies as safe light sources [2]. Charles Darwin also wrote about the glowing oceans in his travails. E. N. Harvey conducted extensive studies on this phenomenon, and wrote the first detailed account of all natural bioluminescent forms. In biochemical terms, the phenomenon of bioluminescence occurs due to an interaction of a substrate luciferin with an enzyme luciferase. Shimomura et al. were the first to obtain crystalline luciferin from the sea firefly *Vargula hilgendorfii* [3].

In this chapter, we explore the origins of bioluminescence in nature, its distribution, and the many ecological roles that it plays. Furthermore, the harnessing of this phenomenon for human use and the future prospects have also been discussed in brief.

IntechOpen

2. The evolution of bioluminescence

Since bioluminescence has proven to be an energy-expensive process, the evolution of bioluminescence in nature must be of some ecological or biological significance, or must offer some evolutionary advantage to the organism. This is certainly true, because there are multiple incidences of the evolution of bioluminescence, all completely independent from each other, and showing a convergent evolution pattern [4, 5]. This trait is found in multiple species spanning different phyla. Some even show symbiotic association with microbes. All these species use this phenomenon for a diverse range of applications including evasion of predators, luring prey and even attracting mates [6–8].

Since bioluminescence is so widespread in nature, scientists have been speculating the cause of its origin and selection in the first place. The first speculation was made by E. N. Harvey himself, who believed that it had something to do with respiratory chain proteins, some of which may have had fluorescent groups or side chains [9]. Owing to the extensive research that he conducted, his theory gained some attention and credibility. It was, however, soon disproved. Some even state that bioluminescence may have merely evolved as a by-product of other metabolic functions, having no importance of its own. However, the repetitive and independent origins of bioluminescence in nature must mean that this trait does confer a significant evolutionary advantage to the species that exhibits it [10].

One theory, proposed by Seliger et al. in 1993, stated that luciferases were actually a group of mixed function oxygenases [11]. According to him, bioluminescence evolved primarily as a means of intra-specific or inter-specific interaction in the dark, deep sea biome.

Rees et al. conducted an independent study on coelentrazine, which is a marine luciferin [12]. They came to the conclusion that bioluminescence may have evolved as a biochemical pathway, mainly for the disposal of peroxide, superoxide, and other harmful oxygen species produced in the course of metabolism. This may have additionally been favored by the acute absence of illumination in the dark depths of the ocean. Bioluminescence may have undergone natural selection as these species may have progressed deeper in the dark depths of the ocean, where the selective pressure for anti-oxidant defense naturally subsided.

As is clear from the above discussion, there was a unanimous agreement among many that bioluminescence may have evolved in the deep sea ecosystem. Even today, the vast depth of the ocean abounds in various species that exhibit this trait. These may range from microbes like bacteria and dinoflagellates to complex organisms like crustaceans, molluscs, jellyfish, various bony fish, and even cartilaginous fish like sharks [10].

As of today, bioluminescence has many more purposes apart from free radical disposal, like camouflage, counter-illumination, warning colouration, predation or courtship, [10] which have been discussed in further subsections.

3. Distribution

As stated earlier, bioluminescence has emerged independently in nature on multiple occasions. Nearly 700 to 800 genera spanning 13 phyla, including both prokaryotic as well as eukaryotic species, have been reported to exhibit this trait [10, 13]. The evolutionary trends of bioluminescence show exemplary convergent evolution in many cases, because of the almost similar purposes this trait serves in various species, or because of the similarity in the biochemistry of the molecules involved.

Bioluminescent organisms are found in both terrestrial as well as aquatic habitats. However, the aquatic species are exclusively limited to marine ecosystems, and a freshwater bioluminescent system is yet to be reported [10].

For the sake of simplicity, the distribution of this trait has been discussed separately for bacteria, fungi and protists, and higher animals have been discussed separately.

3.1 Bacterial bioluminescence

It is a common belief that bacterial bioluminescent systems were among the first to originate in nature. Bioluminescent bacteria are present in both terrestrial as well as aquatic habitats, and can be found all over the world. In fact, these bacteria can easily be sourced from any tissue or detritus lying on beaches, or even from uncooked seafood [4]. The glowing oceans, which are a spectacular result of these microorganisms, have been described in detail in the travails of Darwin, and can be observed, or rather enjoyed at various locations all over the world.

Bioluminescent bacteria mainly belong to the class *Gammaproteobacteria*, and are confined to three genera, namely *Vibrio*, *Photobacterium* and *Xenorhabdus*. Out of these, *Vibrio* and *Photobacterium* are mostly found in marine ecosystems, whereas *Xenorhabdus* inhabits terrestrial habitats [14]. New strains of bioluminescent bacteria are still being discovered [15]. A remarkable fact about bacterial bioluminescence is that all bacterial bioluminescent systems are exactly alike in terms of biochemistry, i.e., they all rely on flavin mononucleotide (FMN), myristic aldehyde and NADH, and also oxygen [16].

Bioluminescent bacteria may exist as free-living, symbiotic or even pathogenic forms. However, a completely obligate bacterial symbiotic system is yet to be observed in nature [8]. For example, *Vibrio fischeri* has been known to colonize specialized "light organs" [17] in the fish *Monocentris japonicus* [18], and also exhibits mutualistic relationship with Hawaiian squid *Euprymnia scolopes* [10, 14], and various species from the genus *Photobacterium* have been known to exhibit symbiosis with various fish, molluscs, etc. [19] and even cause diseases in some others [8]. However, there has been no genetic alteration in the bacterial genome for the said symbiosis. Though the animals showing the said symbiosis have developed exclusive modifications like light organs, they do not show any endosymbiotic behavior. The development of the said specialized organs may even be influenced by the presence of the symbiotic bacterial population [4]. One hypothesis accounts for the emergence of bioluminescence in bacteria because it promotes such symbiotic behavior, conferring a survival advantage to the microbes [10]. The symbiotic behavior may further be promoted because of the fact that the luminescent machinery of the bacteria is instrumental in getting rid of the reactive oxygen species produced in the host tissue [20]. The symbiotic microbes are obtained externally, and the hosts show some degree of selectivity towards the symbiont [8]. It appears that the host organisms 'choose' the colonizing symbiont according to the availability as per the depth which they inhabit. Furthermore, the said hosts can even dump the symbiont cells in order to keep their population in check [20].

Terrestrial bioluminescent bacteria are rare, and are known to infect nematodes that parasitize glowworm larvae. Upon the death of the larva, predators and scavengers ingest the carcasses, hence dispersing the bacteria as well as the nematode. Other than that, bioluminescent bacteria have been observed to inhabit various depths of the ocean, and are found even in sediments, seawaters, saline lakes, etc.

3.2 Fungal bioluminescence

Of all the bioluminescent systems that have been studied, fungal bioluminescence remains by far the most poorly investigated of them all, even though fungi are the only terrestrial eukaryotes that exhibit bioluminescence, besides animals [10]. This might be owing to the fact that most initial attempts at determining the enzymatic nature of fungal bioluminescence were failures, and have only recently been confirmed successfully [21]. The study of fungal bioluminescence has thus gained sudden prominence [22], and a genetically encodable bioluminescent system for eukaryotes has been developed [23]. Kaskova et al. conducted an extensive study of the fungal bioluminescence and colour modulation mechanisms [24].

Out of all the fungal species that have been documented till date, only about 71 [25] to 80 [26] fungal species have been known to exhibit bioluminescence. All of the said species have been unequally classified into four distinct lineages that are not so closely related [23]. "Honey Mushrooms" of the Armillaria lineage, the causative species for foxfire phenomenon, and the "Jack-o-Lantern Mushrooms" from the Omphalotus lineage are common examples of bioluminescent fungi. The origin of fungal bioluminescence can be attributed to a single evolutionary ancestry, the proof of which has been given by cross-reactions between the luciferins and luciferases of distant lineages to yield light successfully [21].

The purpose behind the emergence of fungal bioluminescence still remains elusive. Speculations have been made by Oliveira et al. that it may serve as a mode of attraction for insects, facilitating entomophilous spore dispersal, as seen in some species of *Neonothopanus* [27]. Furthermore, the same study revealed that there is some semblance of circadian control to make this entire affair more energy efficient by increasing bioluminescence at night. However, this is not true for all fungal species, wherein this trait may simply be a luminous by-product of metabolism, without a definite purpose [28]. The evolutionary feasibility of such cases is yet to be determined.

3.3 Bioluminescence in protists

Among protists, the chief groups that exhibit bioluminescence are Radiolaria (or Radiozoa), and Dinoflagellates, which are both exclusively marine. Both of these are described as follows:

3.3.1 Bioluminescent radiolaria

Among all the radiozoa, only two genera, namely *Collozoum* and *Thalassicola* are known to exhibit bioluminescence. Both of these belong to the order Collodaria, and use coelenterazine as substrate [4].

Bioluminescence has also been reported in some other deep sea species like *Aulosphaera* spp. and *Tuscaridium cygneum* [4].

3.3.2 Bioluminescence in dinoflagellates

Dinoflagellates are a group of cosmopolitan protistan organisms [29] having an ancient evolutionary history, which form one of the most important groups of phytoplankton in the aquatic ecosystems [30]. They are the only photosynthetic organisms that are capable of bioluminescence [30], and are the most dominant contributors to the occurrence of this phenomenon in the upper ocean [31]. Common phenomena like the "Red Tides" and the bioluminescent bays of Jamaica are because of the dramatic increase in the population of *Gonyaulax* and other dinoflagellate species. *Gonyaulax polyedra* is supposedly the most studied dinoflagellate species [20]. Other common bioluminescent genera are *Ceratium*, *Protoperidinium*, *Pyrocystis*, *Noctiluca*, [31] and *Alexandrianum* [29]. There have been inaccurate records of bioluminescent dinoflagellate species in the past, because of the presence of both bioluminescent as well as non-bioluminescent strains belonging to the same species. Difference in the ability has been observed even between cells of the same strain [31].

The chemical structure of dinoflagellate luciferin (sourced from *Pyrocystis lunula*) is remarkably unique [20], similar only to that found in euphausiids (krill). This perhaps is an example of dietary linkage, as krill are known to source their luciferin from the food they consume [4]. Dinoflagellate luciferin is believed to be a derivative of chlorophyll [20]. Unlike most species that are autotrophic in nature, some heterotrophic species even supplement their luciferin synthesis with chlorophyll-rich diets [4].

Dinoflagellates produce bioluminescence with the help of specialized cell organelles called "scintillons", which enable them to glow only in response to shear or physical disturbance/turbulence in the surrounding water [31]. This glow is not persistent, but occurs in brief flashes. The intensity of these flashes may be affected by various factors like exposure to prior illumination, nutritional state of the cell, or even because of a diurnal rhythm [31]. There are evidences of a circadian rhythm that is operational in dinoflagellates, and also photoinhibition of bioluminescence during daytime [29]. The synthesis and destruction of luciferin is not the only method of regulation though; cellular redistribution of luciferin has been reported to be affected by the said circadian rhythm [20]. The intensity of the flashes also differs from species to species. Dinoflagellates prioritize bioluminescence second only to reproduction, to an extent that there have been reports of cannibalism under nutritional stress in order to support bioluminescence [31].

As far as the ecological purpose of bioluminescence in dinoflagellates is concerned, we are still unclear as to why these organisms take such measures to sustain it. The exact ecological context of this trait still remains unclear, maybe because of a lack of *in-situ* studies [29]. Some studies show that the flashes of light have a startling effect on copepods (the prime predators of dinoflagellates), which dart away from the prey [32]. Another speculation, called the "Burglar Alarm" hypothesis, states that the brief flashes produced by the cells upon coming in contact with a grazer (for example, a copepod) in turn attracts a predator of higher trophic level, hence protecting the cell from its own predator. This hypothesis is widely accepted, although there are no sufficient evidences of the same [4]. Furthermore, this hypothesis does not point out any clear advantage to the dinoflagellate [31].

To conclude, bioluminescence in dinoflagellates seems to be a useful but unnecessary evolutionary trait, as an accurate ecological context is yet to be determined [30]. In order to gain more knowledge on the same, coastal blooms can be harnessed as natural laboratories to study dinoflagellate bioluminescence in further detail [29].

4. Bioluminescence in animals: distribution and ecological significance

As it is expected, the complexity of bioluminescence certainly upgrades as we proceed upwards in the tree of life. There are no plants (terrestrial or aquatic) that exhibit bioluminescence. Fungal bioluminescence is rare, and has been discussed in the previous sections. Coming to bioluminescence in animals, there is a strong agreement that the evolution of bioluminescence first occurred in the ocean, as the oceanic ecosystem offers many favorable conditions like optical homogeneity, stability of environment, large areas that are almost or completely perpetually dark and a large diversity of organisms that can engage in a variety of ecological interactions [4]. This, and the fact that both luminous as well as non-luminous prey in the ocean are rich in luciferins ensures that the emergence of bioluminescence in the ocean must have been a comparatively easy process [4, 33, 34]. The phenomenon of bioluminescence is so significant in the oceanic ecosystem, that it serves as the predominant source of illumination in many parts of the ocean [35]. Furthermore, courtships involving bioluminescence have been reported to show higher species accumulation rates than those without bioluminescence [36]. The presence of many independent coelenterazine-mediated bioluminescent systems, nine different phyla to be exact [10], indicates dietary linkage, as coelenterazine is procured by most species mainly through their diet [16]. Bioluminescence is encountered most commonly in the topmost 1 kilometer layer of the ocean, and is doubtlessly the most efficient mode of communication in the oceanic ecosystem [35]. The ability to glow is strongly habitat dependent because of various selection forces described earlier, and it is observed that there is a marked difference in the occurrence of this trait as we go deeper in the ocean [35].

Bioluminescence is also common in the terrestrial ecosystems, though it is nowhere as abundant as in the ocean. Various worms and arthropods are known to exhibit complex behaviors related to this phenomenon. It is clear that bioluminescence has a powerful impact on behavioral and ecosystem dynamics [4].

In this section, bioluminescence has been followed as a trait through various animal phyla, both terrestrial and aquatic, and its ecological significance is simultaneously discussed.

4.1 Bioluminescence in ctenophores

Comb jellies are the phylogenetically the most basic examples of bioluminescence in animals. Many species like *Mnemiopsis* [20, 37] use calcium activated coelenterazine as their bioluminescent substrate [4]. Some species, for example *Beroe forskalii* are known to produce myriad, cascading wave-patterns of intrinsic glow on their bodies, and some even emit a haze of glowing particles to startle the predator as a defensive measure, coupled with an escape response [38]. A majority of pelagic species are likely to exhibit bioluminescence [35]. The photo-proteins involved in bioluminescence in various genera like *Mnemiopsis* and *Beroe* have been studied, and are known to depend on calcium ions for their activity [39, 40].

Many comb jellies like *Pleurobrachia* and some species of the genus *Beroe* also show a startling display of rather colorful lights, in various wavelengths found in the visible spectrum. This was mistakenly believed as bioluminescence in the past. However, the said lights were not actually "produced" in the organism itself, as was evident in some studies [41, 42]. This iridescence was rather found to be a result of refraction of ambient light through the moving combs as the organism swims around [43].

4.2 Bioluminescence in cnidarians

Cnidarians in both pelagic as well as benthic zones, including corals, anemones, hydroids and medusae are known to exhibit bioluminescence. All of them use the luciferin coelenterazine as the substrate for their biochemical pathways (hence the name "coelenterazine"). Most of the pelagic siphonophores encountered show bioluminescence [4, 35]. The most common examples of bioluminescent coelenterates is the shallow-living hydrozoan Crystal Jelly (*Aequorea victoria*), the sea pansy *Renilla* and also the bamboo corals from the pelagic zone [44]. Anatomically, light producing centers, or photocytes, may be clustered or widely scattered all over the body, located around the endodermal layer [20]. The bioluminescent system of *Renilla* has been studied extensively, and attempts have been made to triangulate and engineer the genes from the source into various eukaryotic (plant) systems [45].

Cnidarians use bioluminescence for various defensive, agressive as well as warning purposes. Some jellyfish show glowing wave patterns on their umbrellas, and even emit clouds of glowing particles as a part of their escape response [4]. Siphonophores use bioluminescence to attract prey within reach of their cnidocytes. Some jellyfish are also known to show aposematic glow, which is indicative of distastefulness. Cnidarians can gain a lot from aposematic bioluminescence, as it would not only warn the predators of the unpalatability of the individual, but also protect them from any physical injuries [4]. However, many predator species like leatherback turtles use this to their advantage, and easily locate prey like jellyfish.

4.3 Bioluminescence in annelids

Bioluminescence in annelids has independently emerged in several lineages [46], resulting in a rich taxonomic diversity [36] spanning across 45 different genera in 13 lineages of clitellates and polychaetes [7]. They are found in diverse terrestrial and aquatic habitats all across the globe.

Clitellates are the only terrestrial annelids known, including potworms and earthworms from families Lumbridae [47] and Megascolidae [48]. Most of them emit brief flashes, and secrete a slimy coelomic fluid packed with bioluminescent granules [47, 49] under mechanical, chemical [50] or electrical stimulation. The same trend is seen in benthic species from the family Chaetopteridae [46, 51]. This is basically a form of aposematism or advertisement of distastefulness or toxicity [52], due to which predator species avoid such individuals from a distance [7].

In the marine ecosystems, polychaetes are the predominant annelid species in both pelagic as well as benthic zones [53]. Unlike their terrestrial counterparts, marine annelids show an interesting diversity of adaptations of bioluminescence, which they use for a variety of functions. The swarming behaviors of *Chaetopterus* and *Odontosyllis* spp. [51] and their flashing patterns [54] have been studied in detail. The bioluminescent "bombs" of the deep-sea genus *Swima* are detonated upon the slightest disturbance, facilitating an almost ninja-like distraction while the animal swims to safety [55]. Several members of the family Tomopteridae are known to produce golden yellow light, which is quite rare in aquatic ecosystems [56]. Scale worms (family Polynoidae) emanate flashes when disturbed, and even break off one or more bioluminescent scales or even whole parts of the body [57] as decoys or sacrificial lures for the predator while they flee [46]. Some species even shoot sticky glowing mucus at the predators to hamper their mobility, distracting them while making them even more conspicuous [58]. Arrow worms (Chaetognatha) are also known to adapt similar defensive measures. Light production also wards of symbiotic bacteria that overcrowd the tubules of some annelids [59]. Bioluminescence is also used as a mode of intraspecific communication in annelids [7]. Some members of the families Syllidae and Cirratulidae exhibit bioluminescence as a part of their mating behaviors. Elaborate bioluminescent courtship displays of the genus *Odontosyllis* are even known to align with lunar cycles [52, 60].

4.4 Bioluminescence in molluscs

Bioluminescence in molluscs is represented by many unusual taxa, for example the bivalve *Pholas*, the biochemical machinery of which has been extensively studied. Also, the sea-firefly *Cypridina* is a specimen of significance, as its bioluminescent system was among the first to be studied and analysed in detail [3, 61]. The only bioluminescent organism from freshwater ecosystem, the snail *Latia neritoides*, is also a mollusc [62]. Also, the terrestrial snail *Dyakia striata* is another bioluminescent organism that has been studied in great detail [63, 64]. Also, the snail *Hinea brasiliana* uses flashes of blue light as an aposematic signal to ward off predators [65].

Cephalopods are the prominent representatives of bioluminescent molluscs, and some of these may have been the source behind the fables of the mythical Kraken. Among squids alone, there are about 70 bioluminescent genera, both symbiotic and intrinsic [66]. Most luminescent cephalopods use coelenterazine as substrate for bioluminescence [67]. Squids are almost flamboyant in their exhibition of bioluminescence. *Euprymna* is known to be symbiotic with the bioluminescent bacteria *Vibrio fischerii* to form exclusive light organs [10] which it uses for counter illumination [68]. The vampire squid *Vampyroteuthis* has light organs all over its body, and it even shoots glowing particles from the tips of its tentacles. The squid *Taningia danae* has light organs on the tips of its arms, which it uses for intraspecific communication as well as to lure, stun and baffle prey [69]. Even some octopods are known to use bioluminescence to lure prey into their glowing suckers [4]. Cephalopods are also known to autotomize entire glowing arms as decoys if threatened. Some species of octopus also use bioluminescence in courtship displays.

An interesting fact about sperm whales is that they hunt squid by triggering the burglar alarm mechanism around themselves to attract unsuspecting squids.

4.5 Bioluminescence in insects

Insects are the most predominant terrestrial organisms that exhibit the phenomenon of bioluminescence. A majority of the bioluminescent insects are beetles (Coleoptera), click beetles (Elateridae), glowworms & railroad worms (Phengodidae), and fireflies (Lampyridae) [70]. The biochemical mechanism of luminescence is similar in all of these [71], even though each of them emit a diverse palette of wavelengths [20]. Other insects like lantern flies (Homoptera), springtails (Collembola), etc. also show bioluminescence.

Among springtails, only two families exhibit bioluminescence upon mechanical stimulation. Bioluminescence occurs only during sexual phases, and is crucial for sperm transfer. Lantern flies, for example *Fulgora lanternaria*, emit bright white light when both the sexes fly together [72]. Glowworms and Fungus gnats from the order Diptera show bioluminescence only in the larval stages, where they use their glow to attract prey and snare them in webs [73]. The larvae of *Arachnocampa luminosa* are a prime example of such behavior [74]. Female glowworm pupae also glow to attract males [72].

Click beetles show bioluminescence in all stages of life [75]. In the larval stage, bioluminescence serves as a tool to attract prey, as well as for defense. The pupae also glow when illuminated, and adults use bioluminescence for various functions like defense, mating communication and even general illumination [72]. In glowworms, on the other hand, bioluminescence is only secondary to pheromone-mediated communication. Males are rarely bioluminescent, only in the sexual stages for seductive purposes, whereas larvae and females are very luminescent. The railroad worm *Phrixothrix* is highly aposematic, as its body is lined with bright green glowing patches, while it has red headlights, which is very rare among all animals [70].

Fireflies are among the most studied bioluminescent systems, especially the north American *Photinus pyralis* [76]. All life stages in fireflies are luminescent, and firefly larvae are known to use their glow for defensive purposes [73, 77]. Illumination patterns of fireflies may differ even for different individuals of the same species, and are highly encodable [72, 77]. Fireflies have specialized organs called lanterns in their abdominal segments, which can be controlled by the nervous system [20]. Since bioluminescence in fireflies forms the basis of various complex interspecific as well as intraspecific interactions, visual sensitivity according to the environment, time of activity and other parameters has evolved in parallel [78]. The signaling systems in firefly species are highly encodable, species specific, and crucially timed for maximum efficiency. Synchronous flashes are seen in various species, sometimes in swarms spanning 30 meters [72], producing spectacular displays like the ones at Chaophraya river, Bangkok. The biological significance of such displays are still not understood [73]. Due to the uniqueness of the signaling mechanism, some species have evolved to mimic other species specific signals. For example, female fireflies of the genus *Photuris* mimic the female signal of *Photinus macdermotti* to attract and prey upon their males [72]. Fireflies are also highly distasteful to predators, which is exhibited by their aposematic signals, a necessary counter measure to compensate for their high conspicuousness. Today, fireflies are adversely affected by the growing numbers of artificial lighting systems, which hamper their signaling and even cause direct mortality in some cases [79].

4.6 Bioluminescence in crustaceans

The evolutionary pathway of crustaceans reveals that bioluminescence has emerged multiple times. Many krill (euphausiids) are bioluminescent, showing biochemical pathways similar to diatoms [4]. Sergestids use bioluminescence for counter-illumination purposes. Cypridinids are known to release puffs of bioluminescent particles, and also have elaborate mating behaviors involving bioluminescence [4].

4.7 Bioluminescence in other Arthropods

Few luminous species of centipedes (Chilopoda) and millipedes (Diplopoda eg. *Motyxia*) have also been shown to exhibit bioluminescence [50]. Millipedes are also known to show aposematic signaling as a warning for toxicity [80].

4.8 Bioluminescence in echinoderms

Four out of the five classes of echinoderms, namely Ophiuroidea (brittle stars), Asteroidea (starfishes), Holothuroidea (sea cucumbers) and Crinoidea (sea lilies) are bioluminescent [50]. Echinoderms mostly use coelenterazine dependent bioluminescent systems, although some of them also use a novel photoprotein [4]. Bioluminescence is more commonly exhibited by echinoderms inhabiting deep seas. Many new bioluminescent taxa are still being discovered, and 70 ophiuroid species have been recognized to exhibit bioluminescence till date [81, 82].

4.9 Bioluminescence in tunicates

Many species of tunicates are known to exhibit bioluminescence, though planktonic tunicates are not as frequent exhibitors of the trait as planktonic larvacean Appendicularia. However, it cannot be ascertained accurately because some filter feeders (like *Pyrosoma*) may ingest and trap luminescent microbes and appear to be bioluminescent [50]. Species like *Balanoglossus* (Acorn worms) and *Ptychodera* of the class Enteropneusta are also known to be bioluminescent. Also, the sessile adult *Clavelina miniata* glows green when stimulated.

4.10 Bioluminescence in fish

Among vertebrates, fish are the only taxa that have the ability of bioluminescence. This trait is found in fish inhabiting all the depths of the ocean, but is most frequently encountered in specimens from the deepest recesses of the ocean [6]. Bioluminescence is found in about 1500 species of marine bony fish spanning 43 families in 11 different orders [4, 5, 83], out of which some like the anglerfish, flashlight-fish (*Photoblepharon*) and pony-fish (*Leiognathus*) harbor symbiotic bacteria in discrete, specialized light organs, while others produce glow intrinsically [84]. On the other hand, only a handful of shark species in three families of cartilaginous fish are known to exhibit bioluminescence [83]. Unlike bony fish species, cartilaginous fishes do not rely on symbionts for bioluminescence [85], but use an altogether different, unknown bioluminescent system [86]. Some other species like the midshipman fish *Porichthys* and various lantern-fish obtain their respective luciferins from dietary sources [13].

Fish use the ability of bioluminescence for a variety of applications like communication, evading predators, luring prey. The latter is highly expressed in various taxa inhabiting the deep seas. Various anatomical modifications (like the light organs in various bony fish and the esca of anglerfish) harbor symbiotic bacteria, which enable the fish to use the bacterial emission with ample control on the intensity as well as distribution of the emission [4]. Fish of the order Stomiiformes (like dragon-fish, etc.) have evolved most elaborately arranged photophores, including those emitting red light [4]. Cookie-cutter sharks are interesting examples of both counterillumination and mimicry, as they bait their prey with non-luminescent patches on their bodies that look like small fish.

Bioluminescence may also prove disadvantageous to some species in certain cases. For example, elephant seals follow bioluminescence to track down prey populations. Some studies have shown that seals prefer to hunt in locations where there are more bioluminescent individuals [4].

5. Future prospects

Even though we still need to understand the dynamics and biochemistries of many bioluminescent systems in nature, humans have already begun to put bioluminescence to various applications. Bioluminescent mechanisms have been used in the diagnosis of various pathological conditions in the form of Green Fluorescent Proteins (GFP) [20]. Furthermore, attempts are being made to incorporate bioluminescent systems into plants to supplement illumination [87–89]. However, these prospects are still in their developmental stages, and there are various challenges and issues that need to be tackled.

6. Conclusion

The emergence of bioluminescence in nature has occurred independently on multiple occasions, which certainly means that it confers some significant evolutionary advantage(s) which we are yet to understand fully. This is bolstered by the fact that there are so many species that exhibit this trait, and show a plethora of behavioral, anatomical and ecological trends so as to survive and thrive in various habitats. With a better understanding of these systems and their interactions, we will certainly be able to use this phenomenon to our advantage. However, there are some challenges that keep us from fully exploring certain bioluminescent systems.

For example, the deep sea bioluminescent systems are very hard to access, and thus *in-situ* observations are few and far between. With the advent of new tools and techniques, we shall be able to gain a better insight into the dynamics of these systems.

Acknowledgements

The authors extend their heartfelt gratitude to their affiliation institute for encouraging publishing of such works for the readers of the journal. Authors also acknowledge and thank the editors for the opportunity to publish their work.

Author details

Aditya Srivastava and Kalpna Katiyar*
Department of Biotechnology, Dr. Ambedkar Institute of Technology for Handicapped, Kanpur, India

*Address all correspondence to: kalpna@aith.ac.in

References

[1] Dybas CL. Bioluminescent, biofluorescent species light the way to new biomedical discoveries. Oceanography. 2019;32(4):8-9.

[2] Fordyce W. A History of Coal, Coke and Coal Fields and the Manufacture of Iron in the North of England [Internet]. Graham; 1973. Available from: https://books.google.co.in/books?id=dlM_AQAAIAAJ

[3] Shimomura O, Goto T, Hirata Y. Crystalline Cypridina luciferin. Bull. Chem. Soc. Jpn. The Chemical Society of Japan; 1957;30(8):929-33.

[4] Haddock SHD, Moline MA, Case JF. Bioluminescence in the Sea. Ann. Rev. Mar. Sci. 2010;2(1):443-93.

[5] Davis MP, Sparks JS, Smith WL. Repeated and Widespread Evolution of Bioluminescence in Marine Fishes. PLoS One [Internet]. Public Library of Science; 2016 Jun 8;11(6):e0155154. Available from: https://doi.org/10.1371/journal.pone.0155154

[6] Wainwright PC, Longo SJ. Functional Innovations and the Conquest of the Oceans by Acanthomorph Fishes. Curr. Biol. 2017;27(11):R550-7.

[7] Verdes A, Gruber DF. Glowing Worms: Biological, Chemical, and Functional Diversity of Bioluminescent Annelids. Integr. Comp. Biol. 2017;57(1):18-32.

[8] Labella AM, Arahal DR, Castro D, Lemos ML, Borrego JJ. Revisiting the genus Photobacterium: Taxonomy, ecology and pathogenesis. Int. Microbiol. 2017;20(1):1-10.

[9] Harvey EN. The evolution of bioluminescence and its relation to cell respiration. Proc. Am. Philos. Soc. JSTOR; 1932;71(4):135-41.

[10] Kahlke T, Umbers KDL. Bioluminescence. Curr. Biol. [Internet]. 2016 Apr;26(8):R313-4. Available from: https://linkinghub.elsevier.com/retrieve/pii/S0960982216000464

[11] Seliger HH. Bioluminescence: excited states under cover of darkness. Nav. Res. Rev. OFFICE OF NAVAL RESEARCH; 1993;45:5.

[12] Rees J-F, De Wergifosse B, Noiset O, Dubuisson M, Janssens B, Thompson EM. The origins of marine bioluminescence: turning oxygen defence mechanisms into deep-sea communication tools. J. Exp. Biol. The Company of Biologists Ltd; 1998;201(8):1211-21.

[13] Oba Y, Schultz DT. Eco-evo bioluminescence on land and in the sea. Biolumin. Fundam. Appl. Biotechnol. 1. Springer; 2014;3-36.

[14] B Mahajan G, Rahul Phatak D. The Glowing Bacteria - The Living Micro L.E.Ds. Acta Sci. Microbiol. 2019;2(9):06-8.

[15] Gentile G, De Luca M, Denaro R, La Cono V, Smedile F, Scarfì S, et al. PCR-based detection of bioluminescent microbial populations in Tyrrhenian Sea. Deep Sea Res. Part II Top. Stud. Oceanogr. Elsevier; 2009;56(11-12):763-7.

[16] Fleiss A, Sarkisyan KS. A Brief Review of Bioluminescent Systems. Curr. Genet. [Internet]. Springer Berlin Heidelberg; 2019;65(4):877-82. Available from: http://dx.doi.org/10.1007/s00294-019-00951-5

[17] Dunlap P V, Davis KM, Tomiyama S, Fujino M, Fukui A. Developmental and microbiological analysis of the inception of bioluminescent symbiosis in the marine fish Nuchequula nuchalis (Perciformes: Leiognathidae). Appl.

Environ. Microbiol. Am Soc Microbiol; 2008;74(24):7471-81.

[18] Engebrecht JA, Silverman M. Identification of genes and gene products necessary for bacterial bioluminescence. Proc. Natl. Acad. Sci. U. S. A. 1984;81(13 I):4154-8.

[19] Moreira APB, Duytschaever G, Tonon LAC, Fróes AM, de Oliveira LS, Amado-Filho GM, et al. Photobacterium sanctipauli sp. nov. isolated from bleached *Madracis decactis* (Scleractinia) in the St Peter & St Paul Archipelago, Mid-Atlantic Ridge, Brazil. PeerJ. PeerJ Inc.; 2014;2:e427.

[20] Wilson T, Hastings JW. Bioluminescence. Annu. Rev. Cell Dev. Biol. 1998;197-230.

[21] Oliveira AG, Desjardin DE, Perry BA, Stevani C V. Evidence that a single bioluminescent system is shared by all known bioluminescent fungal lineages. Photochem. Photobiol. Sci. 2012;11(5):848-52.

[22] Strack R. Harnessing fungal bioluminescence. Nat. Methods. 2019;16(2):140.

[23] Kotlobay AA, Sarkisyan KS, Mokrushina YA, Marcet-Houben M, Serebrovskaya EO, Markina NM, et al. Genetically encodable bioluminescent system from fungi. Proc. Natl. Acad. Sci. U. S. A. 2018;115(50):12728-32.

[24] Kaskova ZM, Dörr FA, Petushkov VN, Purtov K V., Tsarkova AS, Rodionova NS, et al. Mechanism and color modulation of fungal bioluminescence. Sci. Adv. [Internet]. 2017 Apr 26;3(4):e1602847. Available from: https://advances.sciencemag.org/lookup/doi/10.1126/sciadv.1602847

[25] Desjardin DE, Perry BA, Lodge DJ, Stevani C V, Nagasawa E. Luminescent Mycena: new and noteworthy species. Mycologia. Taylor & Francis; 2010;102(2):459-77.

[26] Chew ALC, Desjardin DE, Tan Y-S, Musa MY, Sabaratnam V. Bioluminescent fungi from Peninsular Malaysia—a taxonomic and phylogenetic overview. Fungal Divers. Springer; 2015;70(1):149-87.

[27] Oliveira AG, Stevani C V., Waldenmaier HE, Viviani V, Emerson JM, Loros JJ, et al. Circadian control sheds light on fungal bioluminescence. Curr. Biol. [Internet]. Elsevier Ltd; 2015;25(7):964-8. Available from: http://dx.doi.org/10.1016/j.cub.2015.02.021

[28] Weinstein P, Delean S, Wood T, Austin AD. Bioluminescence in the ghost fungus Omphalotus nidiformis does not attract potential spore dispersing insects. IMA Fungus [Internet]. 2016 Dec 11;7(2):229-34. Available from: https://imafungus.biomedcentral.com/articles/10.5598/imafungus.2016.07.02.01

[29] Valiadi M, Iglesias-Rodriguez D. Understanding Bioluminescence in Dinoflagellates—How Far Have We Come? Microorganisms [Internet]. 2013 Sep 5;1(1):3-25. Available from: http://www.mdpi.com/2076-2607/1/1/3

[30] Hackett JD, Anderson DM, Erdner DL, Bhattacharya D. Dinoflagellates: a remarkable evolutionary experiment. Am. J. Bot. [Internet]. 2004 Oct;91(10):1523-34. Available from: http://doi.wiley.com/10.3732/ajb.91.10.1523

[31] Marcinko CLJ, Painter SC, Martin AP, Allen JT. A review of the measurement and modelling of dinoflagellate bioluminescence. Prog. Oceanogr. [Internet]. Elsevier Ltd; 2013 Feb;109:117-29. Available from: http://dx.doi.org/10.1016/j.pocean.2012.10.008

[32] Buskey EJ, Swift E. Behavioral responses of the coastal copepod *Acartia hudsonica* (Pinhey) to simulated dinoflagellate bioluminescence. J. Exp. Mar. Bio. Ecol. Elsevier; 1983;72(1):43-58.

[33] Shimomura O. Presence of coelenterazine in non-bioluminescent marine organisms. Comp. Biochem. Physiol. Part B Comp. Biochem. Elsevier; 1987;86(2):361-3.

[34] Harper RD, Case JF. Disruptive counterillumination and its anti-predatory value in the plainfish midshipman *Porichthys notatus*. Mar. Biol. Springer; 1999;134(3):529-40.

[35] Martini S, Kuhnz L, Mallefet J, Haddock SHD. Distribution and quantification of bioluminescence as an ecological trait in the deep sea benthos. Sci. Rep. 2019;9(1):1-11.

[36] Ellis EA, Oakley TH. High Rates of Species Accumulation in Animals with Bioluminescent Courtship Displays. Curr. Biol. [Internet]. Elsevier Ltd; 2016;26(14):1916-21. Available from: http://dx.doi.org/10.1016/j.cub.2016.05.043

[37] Freeman G, Reynolds GT. The development of bioluminescence in the ctenophore *Mnemiopsis leidyi*. Dev. Biol. [Internet]. 1973 Mar;31(1):61-100. Available from: https://linkinghub.elsevier.com/retrieve/pii/0012160673903217

[38] Widder EA, Greene CH, Youngbluth MJ. Bioluminescence of sound-scattering layers in the Gulf of Maine. J. Plankton Res. Oxford University Press; 1992;14(11):1607-24.

[39] Markova S V., Burakova LP, Golz S, Malikova NP, Frank LA, Vysotski ES. The light-sensitive photoprotein berovin from the bioluminescent ctenophore *Beroe abyssicola*: a novel type of Ca2+−regulated photoprotein. FEBS J. [Internet]. 2012 Mar;279(5):856-70. Available from: http://doi.wiley.com/10.1111/j.1742-4658.2012.08476.x

[40] Ward WW, Seliger HH. Properties of mnemiopsin and berovin, calcium-activated photoproteins from the ctenophores Mnemiopsis species and *Beroe ovata*. Biochemistry [Internet]. 1974 Mar 1;13(7):1500-10. Available from: https://pubs.acs.org/doi/abs/10.1021/bi00704a028

[41] Welch V, Vigneron JP, Lousse V, Parker A. Optical properties of the iridescent organ of the comb-jellyfish Beroë cucumis (Ctenophora). Phys. Rev. E [Internet]. American Physical Society; 2006 Apr 14;73(4):041916. Available from: https://link.aps.org/doi/10.1103/PhysRevE.73.041916

[42] Haddock SHD, Case JF. Not All Ctenophores Are Bioluminescent: Pleurobrachia. Biol. Bull. [Internet]. 1995 Dec;189(3):356-62. Available from: https://www.journals.uchicago.edu/doi/10.2307/1542153

[43] Comb Jelly (Ctenophore) | the Shape of Life | The Story of the Animal Kingdom [Internet]. 2017 [cited 2021 Feb 14]. Available from: https://www.shapeoflife.org/news/featured-creature/2017/11/30/comb-jelly-ctenophore

[44] Etnoyer PJ. A new species of Isidella bamboo coral (Octocorallia: Alcyonacea: Isididae) from northeast Pacific seamounts. Proc. Biol. Soc. Washingt. Biological Society of Washington Smithsonian Institution, PO Box 37012, MRC ...; 2008;121(4):541-53.

[45] Mayerhofer R, Langridge WHR, Cormier MJ, Szalay AA. Expression of recombinant Renilla luciferase in transgenic plants results in high levels of light emission. Plant J. 1995. p. 1031-8.

[46] Shimomura O. Bioluminescence: chemical principles and methods. World Scientific; 2012.

[47] Pes O, Midlik A, Schlaghamersky J, Zitnan M, Taborsky P. A study on bioluminescence and photoluminescence in the earthworm Eisenia lucens. Photochem. Photobiol. Sci. Royal Society of Chemistry; 2016;15(2):175-80.

[48] Rota E, Zalesskaja NT, Rodionova NS, Petushkov VN. Redescription of Fridericia heliota (Annelida, Clitellata: Enchytraeidae), a luminous worm from the Siberian taiga, with a review of bioluminescence in the Oligochaeta. J. Zool. Wiley Online Library; 2003;260(3):291-9.

[49] Petushkov VN, Dubinnyi MA, Tsarkova AS, Rodionova NS, Baranov MS, Kublitski VS, et al. A novel type of luciferin from the Siberian luminous earthworm Fridericia heliota: structure elucidation by spectral studies and total synthesis. Angew. Chemie. Wiley Online Library; 2014;126(22): 5672-4.

[50] Oba Y, Stevani C V., Oliveira AG, Tsarkova AS, Chepurnykh T V., Yampolsky I V. Selected Least Studied but not Forgotten Bioluminescent Systems. Photochem. Photobiol. 2017;93(2):405-15.

[51] Deheyn DD, Enzor LA, Dubowitz A, Urbach JS, Blair D. Optical and physicochemical characterization of the luminous mucous secreted by the marine worm Chaetopterus sp. Physiol. Biochem. Zool. University of Chicago Press Chicago, IL; 2013;86(6):702-15.

[52] Gaston GR, Hall J. Lunar periodicity and bioluminescence of swarming Odontosyllis luminosa (Polychaeta: Syllidae) in Belize. Gulf Caribb. Res. 2000;12(1):47-51.

[53] Francis WR, Powers ML, Haddock SHD. Bioluminescence spectra from three deep-sea polychaete worms. Mar. Biol. Springer; 2016;163(12):1-7.

[54] Deheyn DD, Latz MI. Internal and secreted bioluminescence of the marine polychaete *Odontosyllis phosphorea* (Syllidae). Invertebr. Biol. Wiley Online Library; 2009;128(1):31-45.

[55] Osborn KJ, Haddock SHD, Rouse GW. Swima (Annelida, Acrocirridae), holopelagic worms from the deep Pacific. Zool. J. Linn. Soc. Oxford University Press; 2011;163(3):663-78.

[56] Gouveneaux A, Mallefet J. Physiological control of bioluminescence in a deep-sea planktonic worm, *Tomopteris helgolandica*. J. Exp. Biol. The Company of Biologists Ltd; 2013;216(22):4285-9.

[57] Zörner SA, Fischer A. The spatial pattern of bioluminescent flashes in the polychaete *Eusyllis blomstrandi* (Annelida). Helgol. Mar. Res. BioMed Central; 2007;61(1):55-66.

[58] Rawat R, Deheyn DD. Evidence that ferritin is associated with light production in the mucus of the marine worm Chaetopterus. Sci. Rep. Nature Publishing Group; 2016;6(1):1-14.

[59] Morin JG. Coastal bioluminescence: patterns and functions. Bull. Mar. Sci. University of Miami-Rosenstiel School of Marine and Atmospheric Science; 1983;33(4):787-817.

[60] Schultz DT, Kotlobay AA, Ziganshin R, Bannikov A, Markina NM, Chepurnyh T V., et al. Luciferase of the Japanese syllid polychaete Odontosyllis umdecimdonta. Biochem. Biophys. Res. Commun. [Internet]. Elsevier Ltd; 2018;502(3):318-23. Available from: https://doi.org/10.1016/j.bbrc.2018.05.135

[61] Kaskova ZM, Tsarkova AS, Yampolsky I V. 1001 lights: Luciferins, luciferases, their mechanisms of action and applications in chemical analysis, biology and medicine. Chem. Soc. Rev. [Internet]. Royal Society of Chemistry; 2016;45(21):6048-77. Available from: http://dx.doi.org/10.1039/C6CS00296J

[62] Ohmiya Y, Kojima S, Nakamura M, Niwa H. Bioluminescence in the limpet-like snail, Latia neritoides. Bull. Chem. Soc. Jpn. The Chemical Society of Japan; 2005;78(7):1197-205.

[63] Isobe M, Uyakul D, Goto T, Counsilman JJ. Dyakia bioluminescence—1. Bioluminescence and fluorescence spectra of the land snail, *D. striata*. J. Biolumin. Chemilumin. Wiley Online Library; 1988;2(2):73-9.

[64] Copeland J, Daston MM. Bioluminescence in the terrestrial snail Dyakia (Quantula) striata. Malacologia. 1989;30(1-2):317-24.

[65] Deheyn DD, Wilson NG. Bioluminescent signals spatially amplified by wavelength-specific diffusion through the shell of a marine snail. Proceedings. Biol. Sci. [Internet]. 2010/12/15. The Royal Society; 2011 Jul 22;278(1715):2112-21. Available from: https://pubmed.ncbi.nlm.nih.gov/21159673

[66] Nyholm S V, Stewart JJ, Ruby EG, McFall-Ngai MJ. Recognition between symbiotic Vibrio fischeri and the haemocytes of *Euprymna scolopes*. Environ. Microbiol. Wiley Online Library; 2009;11(2):483-93.

[67] Isobe M, Kuse M, Tani N, Fujii T, Matsuda T. Cysteine-390 is the binding site of luminous substance with symplectin, a photoprotein from Okinawan squid, *Symplectoteuthis oualaniensis*. Proc. Japan Acad. Ser. B. The Japan Academy; 2008;84(9):386-92.

[68] Tong D, Rozas NS, Oakley TH, Mitchell J, Colley NJ, McFall-Ngai MJ. Evidence for light perception in a bioluminescent organ. Proc. Natl. Acad. Sci. U. S. A. [Internet]. 2009/06/09. National Academy of Sciences; 2009 Jun 16;106(24):9836-41. Available from: https://pubmed.ncbi.nlm.nih.gov/19509343

[69] Kubodera T, Koyama Y, Mori K. Observations of wild hunting behaviour and bioluminescence of a large deep-sea, eight-armed squid, *Taningia danae*. Proc. R. Soc. B Biol. Sci. The Royal Society London; 2007;274(1613):1029-34.

[70] Viviani VR, Bechara EJH. Bioluminescence and biological aspects of Brazilian railroad-worms (Coleoptera: Phengodidae). Ann. Entomol. Soc. Am. Oxford University Press Oxford, UK; 1997;90(3):389-98.

[71] Wood K V. The chemical mechanism and evolutionary development of beetle bioluminescence. Photochem. Photobiol. Wiley Online Library; 1995;62(4):662-73.

[72] Hoffmann KH. Environmental Aspects of Insect Bioluminescence. Environ. Physiol. Biochem. Insects [Internet]. Berlin, Heidelberg: Springer Berlin Heidelberg; 1984. p. 225-44. Available from: http://link.springer.com/10.1007/978-3-642-70020-0_9

[73] Lloyd JE. Bioluminescence and Communication in Insects. Annu. Rev. Entomol. [Internet]. 1983 Jan;28(1):131-60. Available from: http://www.annualreviews.org/doi/10.1146/annurev.en.28.010183.001023

[74] Meyer-Rochow VB. Glowworms: a review of Arachnocampa spp. and kin. Lumin. J. Biol. Chem. Lumin. Wiley Online Library; 2007;22(3):251-65.

[75] Day JC, Tisi LC, Bailey MJ. Evolution of beetle bioluminescence: the

origin of beetle luciferin. Lumin. J. Biol. Chem. Lumin. Wiley Online Library; 2004;19(1):8-20.

[76] Marques SM, Esteves Da Silva JCG. Firefly bioluminescence: A mechanistic approach of luciferase catalyzed reactions. IUBMB Life. 2009;61(1):6-17.

[77] McElroy WD, Seliger HH, DeLuca M. Insect bioluminescence. Physiol. Insecta. Elsevier; 1974. p. 411-60.

[78] LALL AB, SELIGER HH, BIGGLEY WH, LLOYD JE. Ecology of Colors of Firefly Bioluminescence. Science (80-.). [Internet]. 1980 Oct 31;210(4469):560-2. Available from: https://www.sciencemag.org/lookup/doi/10.1126/science.210.4469.560

[79] Hagen O, Santos RM, Schlindwein MN, Viviani VR. Artificial Night Lighting Reduces Firefly (Coleoptera: Lampyridae) Occurrence in Sorocaba, Brazil. Adv. Entomol. [Internet]. 2015;03(01):24-32. Available from: http://www.scirp.org/journal/doi.aspx?DOI=10.4236/ae.2015.31004

[80] Marek P, Papaj D, Yeager J, Molina S, Moore W. Bioluminescent aposematism in millipedes. Curr. Biol. Elsevier; 2011;21(18):R680-1.

[81] Mallefet J. Echinoderm bioluminescence: where, how and why do so many ophiuroids glow? Biolumin. Focus - A Collect. Illum. Essays. Research Singpost; 2009. p. 67-83.

[82] Jones A, Mallefet J. Study of the luminescence in the black brittle-star *Ophiocomina nigra*: toward a new pattern of light emission in ophiuroids. Zoosymposia. 2012;7(1):139-45.

[83] Paitio J, Oba Y, Meyer-Rochow VB. Bioluminescent Fishes and their Eyes. Lumin. - An Outlook Phenom. their Appl. [Internet]. InTech; 2016. Available from: http://www.intechopen.com/books/luminescence-an-outlook-on-the-phenomena-and-their-applications/bioluminescent-fishes-and-their-eyes

[84] Herring PJ, Campbell AK, Maddock L, Whitfield M. Light and Life in the Sea. Cambridge University Press; 1990.

[85] Renwart M, Delroisse J, Claes JM, Mallefet J. Ultrastructural organization of lantern shark (*Etmopterus spinax* Linnaeus, 1758) photophores. Zoomorphology. Springer; 2014;133(4):405-16.

[86] Renwart M, Mallefet J. First study of the chemistry of the luminous system in a deep-sea shark, *Etmopterus spinax* Linnaeus, 1758 (Chondrichthyes: Etmopteridae). J. Exp. Mar. Bio. Ecol. Elsevier; 2013;448:214-9.

[87] Mitiouchkina T, Mishin AS, Gonzalez Somermeyer L, Markina NM, Chepurnyh T V, Guglya EB, et al. Plants with self-sustained luminescence. bioRxiv [Internet]. 2019 Jan 1;809376. Available from: http://biorxiv.org/content/early/2019/10/18/809376.abstract

[88] Krichevsky A, Meyers B, Vainstein A, Maliga P, Citovsky V. Autoluminescent plants. PLoS One. 2010;5(11):1-6.

[89] Kwak SY, Giraldo JP, Wong MH, Koman VB, Lew TTS, Ell J, et al. A Nanobionic Light-Emitting Plant. Nano Lett. 2017;17(12):7951-61.